REFINERY POLICY IN THE 1980s: SECURITY, ECONOMICS AND EQUITY

Bettina Silber
Clarice R. Feldman
Americans For Energy Independence

Contents

Foreword

The United States is the only major industrialized nation which cannot refine its own needs. At the present time U.S. refineries are operating at capacity. Elsewhere in the world there is excess capacity, and by the middle of this decade, new refining capacity is expected in the Middle East and North Africa.

Unless new domestic refining capacity is developed, we will import relatively less crude petroleum and more petroleum products.

We are at a critical crossroads. Our domestic refineries are primarily geared to process light, sweet oils which are rapidly diminishing and becoming more valuable. The recently discovered finds here and abroad are predominantly heavy, sour crudes which produce a greater percentage of residual fuels, used for electrical generation and in industrial processes. To gear our refining capacity to handle what is coming on line and to meet important environmental requirements, present refineries would need retrofitting, capacity enlargement and upgrading for efficient operation. In addition, new refineries would have to be sited throughout the country consistent with regional demands.

The substantial cost involved in building up U.S. refinery capacity must be weighed against lost employment opportunities, a detrimental balance of payments and resulting inflation, and the security risk of a policy which fosters dependence on foreign refineries.

Does it matter whether petroleum products for domestic consumption are refined at home or abroad? Can we afford to trade dependence on foreign crude oil for dependence on foreign refined products?

On February 4, 1980, Americans For Energy Independence held a conference in Washington, D.C. to address these questions and the underlying national security, economic and equity considerations. Although refinery policy should be an integral part of national energy strategy, it has been the neglected stepchild of energy policy and debate. The present volume, the edited proceedings of the AFEI conference, is an effort to focus public attention on and contribute to greater public understanding of this important topic.

The conference participants were not paid for their efforts. They came because they share our belief that the topic is important and studied debate critical to its sound resolution.

Others who did not participate contributed their insights and time in planning and organizing the conference that resulted in this volume. We gratefully acknowledge their contributions.

For her valuable support in managing the conference and preparing the proceedings for publication, we would like to single out Catherine Piper of the AFEI staff.

Bettina Silber
Clarice R. Feldman
Editors

About the Participants

Chairman

Clarice R. Feldman, General Counsel, Americans For Energy Independence.

Speakers

The Honorable J. Bennett Johnston (D-La.), Chairman, Subcommittee on Energy Regulation, Senate Committee on Energy and Natural Resources.

Hank Banta, Staff Counsel, Senate Committee on the Judiciary. He specializes in energy and antitrust issues and formerly was a member of the Federal Trade Commission staff.

Herschel F. Clesner, Chief Counsel, Subcommittee on Commerce, Consumer and Monetary Affairs, House Committee on Government Operations. He has had 30 years of government service in the Executive Branch, the Senate and the House of Representatives.

Frank Collins, Consultant on Energy and Environment, Oil, Chemical and Atomic Workers International Union, AFL-CIO. He was professor of Physical and Environmental Chemistry at Brooklyn Polytechnic Institute from 1948 to 1976 and has written extensively on science and public affairs.

George Horwich, Senior Economist, Office of Oil Policy, Department of Energy. He is presently on leave from Purdue University where he served as Chairman of the Economics Department. He has also been affiliated with the Brookings Institution and the National Bureau of Economic Research and is the author of numerous works on monetary and general economic policy.

William H. Magee, Vice President, Planning and Control, ARCO Petroleum Products Company. Since 1960 he has held various positions within Sinclair Oil Corporation and ARCO, including Controller, Products Division, and Manager, Planning and Analysis.

Clement B. Malin, Director, Energy Regulations, Strategic Planning Department, Texaco, Inc. His responsibilities involve energy policy and regulatory relationships between government at all levels and the energy industry. He served as Deputy and then Assistant Administrator of the International Affairs Office of the Federal Energy Administration from 1974 to 1977, after 14 years employment with Mobil Oil Corporation.

Stephen E. McGregor, Deputy Assistant Secretary for Oil and Gas Policy, Department of Energy. He is responsible for domestic and international oil and natural gas policy concerns, emergency contingency planning and policy regarding competition within the energy industry. He was previously Director of the Office of Oil and Gas Policy at the Department of Energy. His other experience includes service with the Federal Power Commission and on Senate and House staffs.

James A. O'Neill, Jr., President and Director of Ingram Corporation. He has had 20 years of technical and executive experience in the refining industry. He was the first President of the Energy Company of Louisiana (ECOL), the largest grassroots refinery in the United States.

G. Henry M. Schuler, Member of the Board of Conant and Associates, Ltd., a Washington-based consulting firm, and Director of the company's International Committee of Advisors. In addition to experience as a foreign service officer in Libya and Italy, he has been responsible for international aspects of energy development for a number of companies.

Part I

National Security Implications of Increased Reliance on Petroleum Product Imports

Presentation by G. Henry M. Schuler

At long last the nation is beginning to consider the magnitude of the national security risks which are created by dependence upon foreign oil. There is growing recognition of the incomparable peacetime threat to a strong and expanding national economy, to freedom of action in foreign affairs, to military preparedness and to our international financial standing. In effect, the sovereignty and survival of the United States as we know them depend upon efforts to restore a greater measure of energy self-sufficiency.

Unfortunately, the nation has failed fully to appreciate that dependence upon foreign oil is a multi-faceted problem with many participants rather than confrontation only between the United States and OPEC over crude oil. This failure has led some energy strategists to propose that the United States should increase its reliance upon importation of refined products at precisely the moment that the country is attempting to reduce its reliance upon importation of crude oil. From a national security perspective, it makes absolutely no sense to substitute dependence upon foreign refiners for dependence upon foreign producers.

Americans For Energy Independence is to be commended for being the first organization to bring this issue to public attention. It seems especially appropriate that AFEI is providing this leadership for refining capacity in the only area in which the United States still enjoys independence. We must not let that independence slip away.

I welcome this opportunity to discuss an independent study of *The National Security Implications of Increased Reliance upon the Importation of Refined Products* which was completed by Conant and Associates, Ltd., last summer. The study was sponsored by the Domestic

Refiners Group, an informal grouping of labor unions, industry associations and refining companies which operate in the United States rather than overseas. Conant and Associates is a Washington-based consulting firm founded in 1976 by Mr. Melvin A. Conant, the former head of international operations at FEA, to provide analysis of the political, economic and strategic factors involved in access to energy. Other studies prepared by the firm have included: *Geopolitics of Energy*, prepared for the Department of Defense; *Access to Oil: The United States Relationship with Saudi Arabia and Iran*, prepared for the Senate Committee on Energy and Natural Resources; and *Soviet Natural Gas: Options and Priorities for Soviet Energy Export Policy and Strategic Implications for the West*, which was prepared for The Rockefeller Foundation.

I was the principal author of the study. My own background as an oil industry executive located overseas included participation in the ill-fated negotiations with OPEC in the early 1970s. At the invitation of Senator Frank Church, I testified about that "debacle" to the Multinational Corporations Subcommittee of the Senate Committee on Foreign Relations in January 1974.

The six months that have elapsed since completion of the study have only served to heighten the risks which it addressed. U.S. dependence upon sweet Libyan crude oil has fed the enormous price increases which continue to rock the industrial world. The agony of our allies in Europe and Japan as they try to balance their oil interests against our request for support on the hostage issue serves to heighten our awareness of the risks of oil dependence. The recent conclusion of arrangements whereby Arab producers will process oil in Europe's refineries demonstrates that OPEC's interest in product exports is immediate, not a "pipe dream" down the road. Therefore, it is especially important that we consider national security as the paramount issue in determining U.S. refining policy.

There is substantial evidence that reliance upon foreign refiners will damage efforts to reduce crude oil dependence as well as create new problems and involve additional foreign suppliers in meeting the nation's petroleum requirements. As a result, U.S. national security will face grave new threats, including the following:

• Motivated by a desire to consolidate their control over the world oil trade, the crude oil producers of the Middle East and North Africa have plans to construct or otherwise control five or six million barrels per day of export refining capacity by 1985. Their goal is to extend their indispensability to refining as a hedge against the day that determined consuming countries develop alternative refinery feedstocks from shale oil, tar sands, coal liquids and heavy oil. If they proceed with these plans, particularly those involving joint ventures with the American members of the Seven Sisters, they will expect the United States to import large volumes of refined products in newly constructed tankers which they will also control. This will eliminate the ability of the United States and other consuming countries to juggle world crude oil supplies as they did in 1973 to limit the effectiveness of politically-motivated embargoes. As a result, Arab control over U.S. petroleum supplies will be increased and the danger of a military confrontation will be heightened.

• European refiners and governments are anxious to increase their product exports to the United States in order to utilize the surplus distillation capacity which is seriously undermining the financial viability of the European industry. If the United States were to increase product imports from Europe, that same excess capacity could soon undermine the U.S. refining industry with grave national security consequences. The European capability to meet U.S. demand for gasoline—especially unleaded gasoline—and for low-sulfur fuel oil is very slim. Because supplies would inevitably be subject to interruption in the event of local shortages, the United States would be directly affected by disputes between European countries and oil producers (such as that which led to the recent Nigerian nationalization of British Petroleum) and by differences between the United States and Europe over energy policy and the Middle East. As a result, reliance upon European refineries would create additional foreign policy constraints. This is especially true if—as seems likely—there is a significant increase in processing deals between Arab producers and refiners based in Europe.

• There is also excess refining capacity in the Caribbean and even plans to build new island refineries to serve the U.S. market. Although these refineries are closer to the U.S. military umbrella in time of hostilities, they are located in an area which is becoming increasingly radicalized by Cuban subversion, unfulfilled expectations and racial disputes. These tensions have led to isolated incidents of refinery sabotage and to nationalizations, which could increase at the expense of U.S. consumers dependent upon Caribbean product exports.

The national security advantages to be gained from continuing the historic policy of relying on domestic refineries to supply close to 90% of U.S. product demand are clear, but some would argue that they are offset by the cost advantages of importing seemingly "cheap" foreign products. There is no question that cost must be a major consideration as the nation battles double digit inflation, but analysis reveals that any purported savings will be as transient as those which were argued to support the importation of "cheap" crude oil 10 years ago.

• Although Middle Eastern governments may subsidize their product exports in order to gain initial access to the U.S. market, they will raise prices to a level higher than that which would have prevailed for U.S. refined products as soon as the United States has become vulnerable. This result is inevitable given the higher real cost of refining in remote areas and the much higher cost of transporting product as compared to crude oil. Moreover, it is a virtual certainty that Middle Eastern export refiners will form a cartel to exploit their control and to manipulate prices for political purposes. It would be the height of irresponsibility for policymakers and legislators to ignore the crude oil lesson and force American consumers to relearn it with respect to products.

• Although European prices have been depressed in recent years, they have escalated rapidly in the aftermath of the Iranian Revolution. This has been especially true of the "spot market" upon which U.S. importers would have to depend unless they were willing to make long-term financial commitments in upgraded capacity—in which case any price advantages would disappear. Nor should we ignore the degree of protectionism, favoritism and cartelization afforded to European refiners by many of those governments which seek access to the U.S. market under the banner of "free trade."

• Caribbean prices are also depressed by surplus capacity, but the loyalties of "distress sellers" proved to be as transient as their prices when desperately needed products followed higher prices to Europe instead of to historic U.S. markets during the winter of 1978–79. Moreover, history has demonstrated that importation of "cheap" residual fuel oil did great long-term damage by contributing heavily to the debilitation of the U.S. coal industry and to the lack of development of clean burning technologies. It would be the height of folly to perpetuate this error as the nation struggles to utilize its massive coal resources.

The threat of increased product imports is very real. Most estimates of 1985 imports forecast a doubling of current imports to about four million barrels per day, approximately 20% of total U.S. demand, but some forecasts are even higher. The ability of foreign refiners to capture a larger share of the U.S. market reflects lower operating costs permitted by a lack of commitment to such U.S. national goals as clean air, preservation of coastal ecology, worker health and safety and a strong American merchant Navy. These reduced operating costs are buttressed in some instances by foreign tax concessions, subsidies, market sharing arrangements and financial support.

Since 1972, these advantages have been offset by U.S. price controls on domestically produced crude oil, which provided U.S. refiners with a lower average crude oil acquisition cost than their foreign competitors. The phasing-out of crude oil price controls—while highly desirable as an instrument to encourage production, conservation and conversion—forces policymakers to re-examine that competitive situation in the refining industry.

If foreign refiners are allowed unlimited access to the United States, the American refining industry will suffer from declining utilization rates, reduced profit margins and inadequate investment in conversion and expansion. This will inhibit efforts to diversify refinery feedstocks and otherwise exacerbate the existing problems of crude oil dependence:

• Efforts to expand production of heavy crude oil from Alaska, California, the Naval Petroleum Reserves, Canada's Lloydminster and Cold Lake deposits, Mexico's Reforma Field and Venezuela's Orinoco Oil Belt must be accompanied by massive investments to convert U.S. refineries which were originally designed to handle light crude oil.

• The United States will remain especially vulnerable to supply interruptions and price increases from OPEC's most hawkish members—Libya, Algeria and Nigeria—unless over 50% of American refineries are able to invest in conversion from "sweet" to "sour" crude oil feedstocks.

• Efforts to develop tar sands, oil shales and coal liquids require parallel efforts to equip U.S. refineries to handle the bitumen, kerogen and other synthetics which are to be substituted for crude oil.

Legislators and policymakers must focus on these long-range goals in order to avoid the siren song of foreign refiners promising "cheap" products. Imports must be limited or subjected to a fee which is large enough to offset the additional costs which U.S. refiners incur to achieve desired national goals. It would be entirely reasonable to impose a national security fee on imports. The fee itself would increase government revenues, and it would also generate a national security premium for the domestic refining industry by allowing it to charge prices which reflect the higher operating costs which foreign refiners escape by operating overseas. This approach would recognize that domestic refiners provide the nation and the consumer with a benefit which ought to be compensated for, at least to the extent that domestic refiners incur additional costs to provide other benefits to the nation and its inhabitants.

A firm public commitment to this revived policy of reliance upon a strong domestic refining industry to meet 90% of U.S. product demand would send a signal to the entire world that new foreign capacity should not be constructed to serve the U.S. market. While this signal may not be popular in some producing countries, a far more serious confrontation will develop if Middle Eastern and North African countries proceed with their plans to build export refineries only to find that they are subsequently barred from the U.S. market.

Although the fee should be applied to all imports in the first instance, limited and selected relief might be used as a tool in negotiating overall energy arrangements with neighbors and Western Hemisphere countries.

In summation, let me emphasize that we have applied a test of national security which is as broad as the national debate over energy legislation. The cost of avoiding reliance upon foreign refineries must be weighed against the numerous energy goals which the Administration and Congress have established for the United States in the coming years. They are the essence of national security, for each is critically important to the safety and well-being of every American citizen.

Having thoroughly considered the subject, we are convinced that the evidence indicates that increased reliance on imported products threatens each of the goals in a manner which goes beyond the threat which already exists from crude oil imports:

- The *availability of adequate supplies* to meet normal U.S. demand will be subject to foreign investment decisions, export policies and demand forecasts which do not give priority to the interests of American consumers.

- Foreign *pricing policies* will be exploitative, unpredictable, politically manipulated, cartelized and ultimately reflective of true costs which are higher than product refined in the United States.

- *Competition* within the U.S. oil industry will be affected by allowing increased market share to the largest multinationals, by the existence of refining cartels in Europe and the producing states, by special crude oil ties and by the possibility of predatory competition from foreign governments which are beyond the reach of U.S. antitrust laws.

• The nation's *balance of payments* and the *dollar* will be more weakened by product imports than by crude oil imports because of cost differences, increased foreign investment and the exchange rate applicable to U.S.-European transactions.

• The threat of *supply interruption* is greatly increased because reliance on foreign refineries involves two additional sets of foreign relationships (refiner-producer and refiner-U.S.), raises the vulnerability of installations to terrorist attack or sabotage and limits the flexibility of U.S. refiners during a crude oil embargo.

• Reliance on temporarily "cheap" imports masks the signal of real petroleum costs which is required to encourage *conservation, conversion* and *development of alternative* energy supplies.

• The likely failure of foreign refiners to meet U.S. demand for unleaded gasoline and low-sulfur fuel oil will result in pressures to relax *environmental* restrictions.

• The nation's *military preparedness* is weakened by product shortages during peacetime, and its *national defense capability* is stretched during wartime by the need to protect multiple sets of tanker routes rather than just the principal crude oil passages.

• Freedom of action of *foreign affairs* will be further restricted by providing economic leverage to the governments of the refining countries, by creating new "hostage investments" in the Middle East and by increasing U.S. identification with particular ruling regimes.

The national security implications are clear. A decision in favor of strengthening the domestic refining industry is required in order to avoid a new series of threats and to avoid exacerbating the existing crude oil threat.

Presentation by Hank Banta

I will start off by saying it is a fine idea to discuss refinery policy, and Americans For Energy Independence is to be congratulated for undertaking such an issue. It has probably occurred to anyone who has given any serious thought to petroleum policy that this country could use a coherent, well-thought-out policy toward the refinery industry. There have been moments when I personally have thought a bad policy would be better than no policy at all.

Some may remember that the first National Energy Plan did not contain a single reference to the refining industry. In fact, it didn't even use the word "refining" at all. I suspect that the authors thought that crude oil somehow got mysteriously transformed into product. In practice, all we had for a policy was the perpetuation of the entitlements system which was administered without any clearly stated goal.

The absence of any formally articulated policy did have its advantages—at least to the policymakers. It created, or at least made possible, what I suspect was the largest off-budget welfare program in modern history. All energy policymakers had to do was take the huge sum of money which represented the difference between controlled oil prices and uncontrolled prices and distribute it to such worthy causes as small refiners, Alaskan North Slope partners, et cetera. It had one unintended effect, which is a large part of the subject of our discussion today, and that is that it created a level of protection against imports.

Now, as crude oil decontrol is becoming a reality, we find ourselves with the disappearance of the only refinery policy we had, even though it was an accidental one, and recent debate over what we are to do has become quite noisy.

I have to say at the outset I am a bit pessimistic about the possibility of developing a national public policy in this area for two reasons:

First, the issue is genuinely difficult. The facts, at least those in the public domain, are inadequate, and conventional economic theory is less help than it ought to be.

Second, the politics are impossible.

The central issue revolves around our import policies. The United States is unique, I believe, among industrial nations, in not being totally self-sufficient in refining capacity. What is more important, as has been pointed out, is that the world refining industry has considerable surplus capacity. The fear is that if the United States were to drop import restrictions or not replace those of the entitlements program we'd see a growth in reliance on foreign refinery capacity. Indeed, given the economics of the refinery industry, the operating costs involved in producing an additional barrel being almost zero, we would see the world's spare capacity being used to inflict substantial harm on the domestic refinery industry.

There is also the fact that domestic environmental restrictions, domestic taxes, labor costs, and shipping costs, as Mr. Schuler has pointed out, put the domestic industry at something of a disadvantage compared to their foreign competitors.

All of this raises a series of complex and interrelated issues for which I don't pretend to have any answers. I can only try to enumerate them.

The first issue that must be dealt with is: Why isn't this simply a classic question of free trade versus protectionism? And although the principles of free trade have been much battered in recent years and have become terribly unfashionable, I think the case for liberal free trade policies still has much to be said for it. I am sure everyone here knows the arguments for free trade: comparative advantage, allocative efficiency, consumer welfare, et cetera.

As a matter of public policy we have been reluctant to abandon these values, and rightly so. I have never come across an industry where there wasn't someone who could make a quite plausible case for protection. The problem is if we permitted the case to be made in every industry, the cumulative effect would be disastrous or at least very expensive.

So I think that, to start with, we cannot set the refining industry outside the context of our general trade policy. We must weigh the specific advantages we get in this industry from a more restrictive policy with what we lose in our ability to promote liberal trade policies elsewhere. I am not arguing that that is a determinative factor; I just think it is a factor.

Of course, oil is not just any industry. Crude oil has already been successfully cartelized. It has been used as a political weapon. It already represents, more than any other commodity, the rise of neomercantilism or, more properly perhaps, old-fashioned mercantilism. Moreover, the presence of the OPEC cartel and the dominance of the major multinational oil companies make it impossible to deal with this industry as any other industry.

But once we get past the free trade-protection argument, and we are able to find a case for treating oil differently, we then find a whole host of other issues.

The first of these issues is the classic American conflict of independents versus major companies. The independent refiners argue that although they represent only a small part of the industry, they have a disproportionate effect on the market in the way of reforms. They claim that they are a serious competitive force in the market and that without them the industry would be hopelessly collusive.

In addition to their competitive behavior, they claim to be more efficient and more innovative. I remember that the president of one large independent refinery once told the Senate Antitrust Subcommittee that his company could spot the major oil companies a dollar a barrel in crude costs and still underprice them at the pump.

The independents-versus-majors question is now, as it has been in the past, a question of access to crude at equal cost. But today the independents are finding themselves faced with a far greater crude cost than their major rivals. The entitlements program has as its purpose the equalizing of the disparity of cost caused by domestic price controls. It was never designed to deal with differences in costs among various foreign crudes. Now these differences have become very great and the independents are becoming very nervous, particularly as they look towards decontrol. They are quick to point out that most of the offshore refineries belong to the majors, and they would benefit from a liberal import policy. The independents are being joined by a number of the domestic integrated refiners, those companies whose integration has in the past depended upon the domestic production. As their domestic production begins to decline, they, too, become more dependent upon foreign crude, and they are beginning to talk a lot more like Ashland than like Exxon.

One of the most interesting documents to come out of the petroleum industry in recent years is the petition which was filed by

Union Oil for entitlements relief before DOE in the latter part of last year.

The competitive issue brings us to one of the real failures of our economic theory. There is a great debate over what the major oil companies are likely to do with the cost advantage they have. The independents are willing to suggest that we will see some really truly predatory behavior. Others claim such scenarios are inconsistent with any theory of profit maximizing on the part of the major companies. All of this points to the fact we have no theory about vertical integration in the industry. DOE is now under congressional instruction to look at the question of what we call, for lack of a better word, cross-subsidization. I think the study is restricted to the subsidization of marketing. Hopefully, if they keep winning their subpoena battles, we will learn something about the whole issue.

If one accepts the thesis about the competitive role of the independent refiner, one becomes stuck with the problem of how to deal with the issue in the context of an import policy on refined product. Obviously, a restriction which would simply raise the price of oil to a point where all could survive comfortably would be an acceptable solution to the independents, but it would be logically inconsistent with the argument that they bring economic benefits to the consuming public.

The question of how to deal with the survival of independent refineries without distributing an enormous amount of public money to the whole industry is a very serious one and I don't pretend to have any answer.

Before leaving the competitive question, I would like to take issue with one point that Mr. Schuler raised, and that is the question of the small refiners.

I don't think one can really make a case for subsidizing inefficiency solely on the grounds that it is inefficient. The demise of the small refiner bias would not break my heart. I guess I can conceive of a program which would continue to subsidize them, but I would find it appalling.

If we leave the purely economic question and move into the question of national security, I think Mr. Schuler has elaborated on that quite adequately, and I don't want to add very much to it except to raise two rather difficult points.

The first is a rather familiar problem to people who have been involved with the refining industry, and that is the question of what is secure and what isn't. The question basically revolves on how we treat the Caribbean and other offshore refineries. How do we develop a policy which deals with imports yet does not inflict harm upon our allies who are economically dependent upon their trade with the United States? It would be a serious mistake to adopt in the name of national security an import policy which damaged the economies of countries friendly to the United States. It should be remembered that the refining industry offshore of the United States has been constructed in response to past American policies. Whether those policies were right or wrong, they were there. They were relied on. Any move which would substantially reduce the ability of friendly countries to export to the United States would create serious internal political problems for them, which in turn would have other national security ramifications.

Mr. Schuler has raised a more fundamental problem: the whole question of whether or not OPEC's moving downstream into refining, perhaps even into marketing, and the extent to which that creates a security problem. I would like to add one other consideration which is often overlooked, and that is that there has never been a successful refinery cartel. Those of us who are on the government side of antitrust are constantly citing the "as is" agreement in the 1920s, when the petroleum industry attempted to form a refinery cartel. The thing we usually leave out is that it failed, it didn't work at all, and they got caught.

They kept exchanging letters and memoranda and complaints with each other. It was inevitable that they'd get caught. As the Department of Energy has found out, when you are trying to freeze prices and market share, you need exceptions and special rules; you need hardship relief and allocation agreements. The industry was practically duplicating the Department of Energy. That is hard to do.

It just flat didn't work. And it didn't work because of the nature of the refining industry. One has the suspicion that if OPEC really tried to run a refinery cartel it couldn't do it. And is it really in our interest to discourage their construction of a refining industry?

Again, I am not claiming that this consideration disposes of the issue. It certainly doesn't. But I think it is a factor that deserves a lot more careful attention than it has been given.

I started out by saying I was pessimistic about our ability to develop a refinery policy. However, I think there is one enormous benefit from the debate, and I think the benefit is accruing to the people who approach the petroleum industry with curiosity. It is appropriate to drag out here that worn-out quote about how when oil men get together they are like cats—you can't tell whether they are making love or fighting.

Today I think it is clear they are fighting. And the benefit is we are learning more from it than we ever learned before. The willingness of members of the oil industry to come in and snitch on each other is absolutely wonderful. It is one of those rare occasions when we outsiders can learn things. I would not like to see us develop a refinery policy before we learn a lot more.

Presentation by George Horwich

The Department of Energy has not yet formulated its position on refinery policy under decontrol. The research reported here was conducted by the Office of Oil Policy as part of the general effort to reach a policy consensus. I will inject some of my personal views into the account of the research program. I will also take this opportunity to summarize and criticize the refinery policy proposal of Conant and Associates, Ltd., of which our co-panelist, G. Henry M. Schuler, is the principal author.

U.S. Refinery Capacity and Petroleum Imports Under Decontrol

It is by now well understood that decontrol of U.S. crude oil prices will lead to some reduction of domestic refinery capacity as crude prices in this country climb to world levels. Smaller, higher-cost refineries, in particular, will be less able to compete once they lose the advantage of lower, controlled crude prices and the small refiner bias under which small refiners are in effect allocated a more-than-proportionate share of the low-cost domestic crude.

As some U.S. refiners lose their competitive advantage, foreign refiners, particularly in the Caribbean and, to some extent in Europe, will gain an advantage. U.S. imports of petroleum products will increase. DOE estimates that domestic capacity will fall and product imports will rise anywhere from 500,000 to one million barrels per day—say, as an average, 750,000 barrels per day.

The rise in product imports does not mean that total petroleum imports will go up. As product imports rise, domestic refinery output falls. U.S. refiners will thus need less crude oil, and imports of crude will fall, barrel for barrel, with the rise in imports of product.

Total petroleum imports not only fail to increase under decontrol, but the total actually goes down. This is because as U.S. prices of both crude oil and oil products rise with the removal of controls, the demand for each falls. We end up with a smaller import total in which products are a higher proportion and crude a lower proportion than before decontrol. The lower total import level is, of course, one of the primary goals of the decontrol program.

The question we now face is whether this rise in petroleum product imports is truly in the national interest. The Administration, through a long process of analysis and soul-searching, concluded nearly a year ago that decontrol was the proper route to take. Having made that decision, it may seem logical to say that firms which cannot then survive without the subsidy of price controls and entitlements should be permitted to go the way of the market. In general, there are certainly scale economies in refining and it is indeed likely that some small refineries in the under-50,000 barrels per day class that do not offer a broad range of products or have particular locational advantages will not outlast crude oil decontrol.

The Office of Oil Policy has completed a study to show what it would cost to protect the domestic refinery industry from the effects of decontrol by means of a tariff on product imports. A tariff causes cheaper foreign goods to be replaced by more expensive domestic ones. We refer to this as the "real resource cost" of the tariff. In the case of a tariff on petroleum products, the real resource cost, at 1978 oil prices, would be several hundred million dollars per year. But, in addition, there would be transfers from consumers to refiners and government of over $6 billion per year. At today's oil prices, those magnitudes would be even higher. Protection for the refinery industry will not be cheap.

But, obviously, economics cannot be the only consideration. There may be national security risks in allowing a certain amount of domestic refining capacity to be replaced by petroleum product imports. These risks may far outweigh the narrow economic costs involved in preventing product imports from rising above their current levels. This question of national security, as it relates to refinery policy under decontrol, was specifically addressed by the Office of Oil Policy.

Petroleum Product Imports and National Security

It is important to understand clearly what the national security risks of a free-market refinery policy are. Once again, it is not a question of our total volume of imports—that, in any case, is already falling and will be lower still under full decontrol. The risks lie in the fact that under free trade and decontrol, domestic refinery capacity will be less and imports of finished petroleum products will be greater than they have been under controls. Imports of crude oil will be lower. Most of the additional product imports will come from the Caribbean and Europe, where there is considerable excess refining capacity.

We introduced this pattern of imports under decontrol into a simulated model of the world oil market. Then, by means of a tariff on product imports, we restored the predecontrol import pattern, in which products are lower and domestic refinery capacity and crude imports are higher. Next we subjected each import scenario—one with a higher product and a lower crude level, and one with fewer products and more crude—to a wide variety of oil supply interruptions originating in various parts of the world.

If there were any significant general advantages to one import scenario over the other under the interruptions, we could not find them. Neither in terms of the stability of total U.S. imports, nor other measures of the costs of interruptions such as price increases or consumer losses, could it be said that it is generally better for the United States to have fewer imports of product and more crude rather than more crude and fewer products. Indeed, it does not make a great deal of difference whether, under free trade, we import more products from the Caribbean and Europe, which are subjected to crude interruptions, or whether, under U.S. refinery protection, we import fewer products and experience cutoffs of crude supplies directly.

It is true that when a product import tariff protects the domestic refinery industry, U.S. product prices are higher and our total petroleum consumption and imports are somewhat lower than under free trade. As a result, our vulnerability to world supply interruptions, as measured by import losses or price increases, is slightly less than under free trade. But the costs of a permanent product import tariff have to be balanced against the slightly reduced losses occurring under occasional oil supply interruptions. In these terms, it is hard to justify the tariff. It would take a loss of several million barrels per day of world crude supplies occurring every year, year in and year out, to pay for the real resource and consumer costs of the tariff. There are far more efficient ways than a product import tariff to reduce the costs of interruptions. One is a tariff on all petroleum imports—crude and product alike. This reduces consumption without incurring costly resource reallocation. Another is a large Strategic Petroleum Reserve.

The reason we do not generally gain by having more, rather than less, refining capacity is that when crude oil interruptions occur, unused refining capacity springs up automatically and immediately throughout the world. This occurs in the United States, as well as in nearby refining centers in Eastern Canada and the Caribbean. However, there would not be any excess refining capacity if the interruption took the form of a shutting down or destruction of refineries themselves. Suppose, under free trade and high product imports from the Caribbean, a natural or other disaster cuts us off from our Caribbean suppliers. In this event, even our Strategic Petroleum Reserve would be of limited value, owing to the absence of simultaneously emerging excess refining capacity.

In answer to this last contingency, we make the following points:

1. The choice is between importing petroleum products from the Caribbean under free trade or, if we protect our domestic refining industry, importing crude oil from OPEC. Clearly OPEC has been a much less reliable source of oil than the product-exporting islands of the Caribbean. And if we should ever reach the point of military involvement, there is much more to be said for intervening in the Caribbean than in the Persian Gulf.

2. In any case, oil supply shortfalls tend, sooner or later, to be spread around the globe, pretty much in proportion to each region's share of world petroleum consumption. This will result from the International Energy Agency sharing plan or powerful natural market forces. This tendency makes most embargoes ineffective and weakens the case for avoiding oil dependence on one part of the world instead of another. In the end, there is one world and one market for oil. As long as we are importing any products from anywhere in the world, a cutoff of product exports from the Caribbean will ultimately cost us our proportionate share, even if our imports from that area are initially zero.

None of these findings should be construed as a judgment on our part that the United States does not need new refineries. We hope and expect that in the future private investors will build small and large facilities as the need arises. We feel it is important that the decision to do so be based on the willingness of private entrepreneurs to risk their own capital. The Department can and should play a catalytic role by trying to speed up the regulatory delays due to environmental standards and the like.

Response to the Conant Report

In the time remaining, I would like to direct my remarks to the broad criticism of decontrol and the policy of free trade made by Mr. Schuler on our panel this morning. As he pointed out, his analysis was based on the larger report on the same subject written by him for Melvin Conant and Associates last summer.[1] Not only does the Conant Report see national security dangers in our importing products from the Caribbean and Europe, but, even more ominously, it sees Arab OPEC constructing refinery capacity and entering the export market on a large scale.

The report makes the following points:

• Crude oil producers in the Middle East are planning to construct export refining capacity of five to six million barrels per day by 1985. The United States will be expected to take a substantial share of these exports at prices which initially will be subsidized by crude oil revenues. Later, prices will be increased as the producers extend their oil cartel to include refinery operations.[2]

• An increase in U.S. product imports from Europe is equally fraught with danger. European refiners would be unable to meet the U.S. demand for unleaded gasoline and low-sulfur fuel oil. Supplies would be unstable because of local disputes and differences between the United States and European countries on policies with regard to energy and the Middle East. There is also the very real "protectionism, favoritism, and cartelization afforded to European refiners by many of these governments."[3]

• Nor does the Caribbean promise to be a secure source of product imports. While the area is under our "military umbrella . . . it is becoming increasingly radicalized by Cuban subversion, unfulfilled expectations and racial disputes."[4]

In view of all this, the Report proposes a protectionist policy. "U.S. refineries, unlike their foreign counterparts, have higher operating costs due to our commitment . . . to clean air, . . . worker health and a strong American merchant Navy."[5] Decontrol should be coupled with a "national security fee on imports" so that domestic refiners can charge a price to cover their higher costs and to provide the benefits of avoiding dependence on foreign suppliers.[6]

I will comment on these points, one by one, though not necessarily in the order presented.

The above statement of the problem, based on the Executive Summary of the Conant Report, is misleading because of what it omits. Although the report acknowledges in various places that a limitation on product imports results in greater imports of crude, that crucial fact is not given the emphasis it deserves. Remember that when refiners in Europe and the Caribbean are shut out from the American market, we do not thereby reduce our total petroleum imports. The protected American refiners will need more crude for their operations. For every barrel of imported product excluded from the U.S. market, a barrel of imported crude takes its place. The choice we face is between buying products from the Caribbean and Europe, which rely largely on OPEC crude, or limiting product imports from the Caribbean and Europe and buying the OPEC crude ourselves.

I do not see significant national security advantages of one alternative over the other. The total distance traveled by the oil is roughly the same, whether Middle East or North African crude is refined in the United States or in ports in the Mediterranean and Caribbean en route to the United States. A security case can in fact be made for letting refiners in Europe or the Caribbean, rather than ourselves, have the pleasure of buying the crude directly from the OPEC producers. We may even gain a degree of latitude in foreign policy as a result of limiting our economic dealings with OPEC.

The economic case for refining the crude at intermediate points in the Mediterranean or Caribbean will speak for itself. Under free trade and decontrol, the large diversified and under-utilized refineries in these areas give every indication of achieving lower costs and greater comparative efficiency than many of our smaller refineries that owe their existence to price controls. In any event, the market will soon tell us if this is an accurate assessment.

The world export-refining industry is highly competitive. Refiners in Europe and the Caribbean, with or without the assistance of their governments, have not been able to gain control of the market—nor will they, as a result of our buying more products from them. Either these refiners now and in the future offer us products as cheaply as they can be obtained elsewhere, and with a reasonable degree of stability, or we will buy them elsewhere. If they do not offer us sufficient quantities of unleaded gasoline or low-sulfur fuel oil, then we will seek alternative supplies; more likely, we will produce these products ourselves, as we have done with unleaded gasoline in the past. The refined product market has very little in common with the crude market, where control over significant scarce supplies is indeed in the hands of a few collusive producers. Refinery capacity is relatively much less scarce and totally uncartelized.

Mr. Schuler's argument that American refiners must be compensated for the higher costs of doing business in America is, of course, the same argument that has been made for every industry, segments of which face competitive pressures from which it seeks protection. American wages have long been the highest in the world. Our ability to compete in world markets has come from efficiency and productivity levels that give American industry a comparative advantage, even while paying the world's highest wages. In recent years American workers have received some of their compensation in the form of healthier working conditions, including a cleaner environment. Mr. Schuler argues that these benefits are of a special character and impose business costs that should be subsidized by taxpayers and consumers-at-large. Under this reasoning, any group of firms whose ability to compete is slipping is a potential candidate for eternal support by a benevolent government.

The statement that OPEC producers are already entering significantly into the refining export market (and presumably would be encouraged to do so by any decline in U.S. refinery output) is not substantiated by DOE's 1979 survey of the world refinery industry.[7] In the Middle East, any additional export refining capacity will appear before 1983 only because of reduced internal consumption by Iran.[8] Beyond 1983, 1.4 million barrels per day of export refining capacity are scheduled to come on stream.[9] But this is 0.5 million less than was forecast in 1978.[10] And more than half of the 1.4 million total, 0.8 million, is in the "study" stage, compared to only 0.3 million so characterized the year before.[11]

Among other OPEC members, contemplated increases in refinery capacity in the next two years total only 0.5 million barrels per day (mainly in North Africa).[12] The general picture is one of OPEC building refineries to keep pace with its own growth and internal requirements. Beyond that, its near-term attempts to enter the refinery export market appear very tenuous, as one would expect in a world that already has significant excess refinery capacity. The claims of Arab and other members of OPEC, frequently overstated in the past, cannot be taken as a basis for U.S. policy in the present.

The Conant argument that a monopolist can extend his power by tying in sales of a good in which he has monopoly power with product sales in which he is a competitor is a common misconception. It is difficult to see what economic advantages would accrue to OPEC producers if they were to tie in refined products with their crude. They would sell more product but less crude, a larger portion of which would be retained as raw material for their new refineries. Sales of their low-cost crude would be replaced by sales of product for which OPEC's comparative advantage is no greater, and probably less, than numerous other countries of the world. OPEC's total sales of petroleum—crude plus product—would be essentially unchanged, its costs higher, its profits lower, and its total leverage over consuming nations no greater than previously.

If, nevertheless, OPEC decides for other reasons to build refinery capacity which would not be to its economic advantage, it will surely not be deterred by a U.S. decision to protect its own refinery industry.

References

[1]G. Henry M. Schuler, *The National Security Implications of Increased Reliance Upon the Importation of Refined Products* (Washington, D.C., Conant and Associates, Ltd.), July 1979; also, the *Executive Summary* to the above.

[2]Schuler, *Executive Summary*, pp. ii and iii.

[3]*Ibid.*, p. ii.

[4]*Ibid.*, p. iii.

[5]*Ibid.*, p. iv.

[6]*Ibid.*, p. vii.

[7]U.S. Department of Energy, *Trends in Refinery Capacity and Utilization, Through June 1979* (Washington, D.C., DOE/RA-0010[79]), September 1979. The Conant Report's projection of 5 to 6 million barrels per day of export refining capacity in the Middle East and North Africa by 1985 is based mainly on assertions by OAPEC Secretary General Ali Attiga (p. 50). The Report also cites estimates by the Shell Co. of 6.5 million barrels per day and by the *Oil and Gas Journal* (April 24, 1978) of 7.5 to 8 million barrels per day (p. 84, n. 15). The source of the Shell estimate is not identified, although the *Oil and Gas Journal* issue cited refers to Shell on p. 49 as estimating "2 million barrels per day under design in the Middle East and North Africa alone, plus 1.5 million barrels per day of potential additions by 1985." The *Journal's* own estimate of refining "construction underway" in the Middle East and North Africa is 0.7 million barrels per day, and of "planned projects," 3.9 million barrels per day (p. 47). No further characterization of these "plans" is given; they are in any case far below the amount attributed to the *Journal* by the Conant Report.

[8]*Trends, op. cit.*, p. 49.

[9]*Trends, op. cit.*, p. 67.

[10]*Trends, op. cit.*, p. 66 and U.S. Department of Energy, *Trends in Refinery Capacity and Utilization, Through June 1978* (Washington, D.C., DOE/RA-0010[78]), September 1978, p. 53.

[11]*Trends, Through June 1979, op. cit.*, p. 66, and *Trends, Through June 1978, op. cit.*, p. 53.

[12]*Trends, Through June 1979, op. cit.*, p. 68.

DISCUSSION

MR. SCHULER: I readily acknowledge that if we don't import refined products, we import crude oil instead. The point I emphasize is that there is a great difference between importing crude oil and importing product. In a number of respects, importing product has a higher cost and provides more detriment to the national security. I pointed out that the cost of imported petroleum products is considerably higher, 40–60% higher in many instances, than that of crude oil. That is because there is the refining charge, there is higher transportation cost of bringing it to this country.

Now, that is not true, as I pointed out, for residual fuel oil. If that is all we are going to import, that economic argument is not as clear. But we are trying to get away from residual fuel oil, and if we fail to build refineries in the country to make the unleaded gasoline and low-sulfur middle distillates we require, we will be importing these products in the future. So as a result we are going to have a worse balance-of-payments drain, for a start, than we have now.

Secondly, from what Mr. Horwich has indicated, he has assumed that we have the same disputes with the oil producers as everybody else does. That is simply not the case.

Canada had a dispute going with the Middle Eastern oil producers within the last several months, when Prime Minister-Elect Clark indicated he was going to move the Canadian Embassy in Israel from Tel Aviv to Jerusalem. The Arabs threatened to cut off Canada's crude oil supplies and a lot more than that. Now, we weren't involved in that dispute. Suppose we were dependent upon Canadian refineries, that we let our own refining industry atrophy to depend on Canadian industries in this regard, and they had their crude oil cut off. If you follow the DOE's scenario, we go to the Persian Gulf and say, "We will buy more crude oil," but we haven't built the new capacity and can't handle the sour crude oils.

And that is not an isolated case. What if we were dependent upon Britain for our refined products? Britain had a dispute going with

Nigeria in the last eight months. Nigeria nationalized the British Petroleum Company and denied imports to the United Kingdom that Britain was previously getting. It was a dispute we were not involved in at that particular point in time.

So we add another element of uncertainty if we have a refiner standing between the crude oil source and ourselves.

Again, the investment decision that needs to be made is: Why should we depend upon European or Caribbean or Canadian refiners to put in capacity to meet our very unique requirements for unleaded gasoline and low-sulfur fuel oil? It just seems to me the incentive is not there when they know if they make that investment we can at any point in time put in that tariff that we might be unwilling to put in now. So they are not going to make those investments to fill our needs.

The only way we can have a secure, reliable, adequate, reasonably priced source of products is to rely on refineries in this country, not refineries that somebody else is going to build and control overseas. I think we have to keep that in mind as we look at this issue.

MR. HORWICH: I hear the siren song of protectionism, and I am still not persuaded by it. I am not inspired by the prospect of trading reliance on Middle Eastern crude for, let us say, Canadian or Caribbean product.

Reasonable people might well favor one as opposed to the other. The point is that under decontrol and free market decision-making American producers will decide where their best interests lie—and I have a lot of confidence in the outcome of that decision. If Canadian refined product doesn't have the particular characteristics we want it to have, including a high degree of reliability and stability, and if European countries are unwilling or unable to send us low-sulfur fuel oil or unleaded gasoline at prices that we can match at home or do better than, then there is no question in my mind where the solution lies. We will produce the stuff at home or buy it where we can get it at the lowest price and the most stable conditions.

I think we ought to let the market sort these decisions out. I am not at all intimidated by the prospect of these enormous costs of transporting the products overseas because I think that if indeed they occur we won't transport them; we will produce the stuff at home. Again, I have enormous faith in the ability of American entrepreneurs to make that decision, not to lock themselves into something they cannot count on or that can be produced at home more cheaply.

MR. BANTA: I'd like to ask both of my fellow panelists how they would deal with the question of disparity in crude costs, which seems to be a matter of enormous concern to virtually everyone in the industry.

MR. HORWICH: Are you talking about the different crude prices paid by large and small refiners?

MR. BANTA: Yes, this is underlying the discomfort the independent refiners are facing and their fear for the future under decontrol.

MR. SCHULER: The only answer that I can see—and I don't know that it is a desirable answer—is that the problem is going to be eliminated by the producing countries. Take the case of Saudi Arabia. They are not going to allow the ARAMCO partners to receive in perpetuity the increment between the world price and the price they get on a buy-back arrangement. The Saudis won't allow that to continue forever. They will simply take it away.

MR. HORWICH: I haven't studied that problem in any detail. People in the Oil Policy Office are looking at it. I don't know to what extent it is a real problem or whether it reflects the extent to which small refiners are already losing their competitive advantage.

MR. BANTA: It has resulted in awesome powers by the Office of Hearings and Appeals, which has undertaken to become the supply and distribution manager of virtually the whole petroleum industry.

MR. SCHULER: It is a very real problem. I just think it is probably a shorter-term problem.

MR. HORWICH: In any event, adjustments to new price levels, to free-market prices instead of controlled prices, are always very painful. The process is bound to have problems.

MR. WILLIAM A. STOOPS (Washington Analysis Corporation): Mr. Schuler, what can you do if the Arabs are expanding their capacity and decide to tie our purchase of crude supplies to purchases of their product?

MR. SCHULER: That is precisely the point that I want to address and have tried to address. Quite frankly, once their refining capacity is built and that investment is made in refining capacity, there will be little or nothing we can do to avoid having products tied to crude oil supplies.

I believe that we can avoid the investments being made in that refining capacity by sending a very clear signal at this time that we are not going to be open to the import of those products. And for once we can create some uncertainty in the mind of OPEC as to whether we will be willing to take their products. If we don't send the signal, if we continue to drift, there will be significant new refining capacity built in the Middle East and North Africa.

I went to a number of sources to determine how much is being built. They have talked, as Mr. Horwich said, about as much as four to six million barrels per day of export refining capacity.

Now, I don't think they are going to build all that. But if we continue to drift and if they continue to assume that they can muscle their way into this market, they will build that refining capacity, and then we will really have a problem because we will have products and crude oil tied to each other. So this is precisely why I think now is the time to send a signal that this capacity should not be built, that the investments they want to put into refining should be put into the search for additional crude oil supplies. There is a lag in exploration in the Middle Eastern states, and there is no place to look for crude oil like where crude oil is. They can very well put their investment into that.

But at the rate we are going now, those refineries will be built. A lot of them will have major American oil companies as partners, coventurers, because the Saudis and others are looking to the coventurer to provide a U.S. market for that product.

MR. STEPHEN GOLDBERG (Office of Senator Bentsen): I have a question that picks up from Mr. Banta's point. He says OPEC has not yet worked out a refinery cartel. But aren't we seeing a situation developing where we have countries like Algeria and Libya raising their prices on sweet crude oil, which is especially important to the U.S. refining industry because so many of the new entrants in the refining business depend on sweet feedstock to run their refineries appropriately? If this happens, what is going to happen to the small refiners when their crude sweet stock is going up very, very much higher than the normal world OPEC price rise?

MR. BANTA: I suppose you've put your finger on one of the irrationalities of domestic refinery policy. We have put literally billions into small refineries. Not one, to my knowledge, can run high-sulfur crude or make unleaded gasoline or do any of the things we need. One has to question the policy of proliferating small refiners that are so irrelevant to meeting any legitimate national need.

The question of the ability of OPEC to run a refinery cartel is to some extent related to their ability to adjust the prices of varying degrees of crude oil. Up until the tight market, I think it is fair to say they have not done that well. One of the things the OPEC nations have never been able to resolve is the quality differentials and transportation differentials. But in any case, running a crude oil cartel is simple compared to running a refinery cartel. If the Seven Sisters couldn't run a refinery cartel, one wonders whether OPEC can. Without some systematic control for crude, I just don't think you can run a refinery cartel.

MR. SCHULER: I agree it's a much more complicated question. But we can't forget that the OPEC refiner cartel is simultaneously the OPEC producer cartel, which gives them a leverage over all the rest of the refiners in the world who are dependent upon that crude oil. And we have seen OPEC extend outside of a simple crude oil cartel. They are moving rapidly toward an LNG cartel; they are moving toward an LPG cartel. They have a good deal in the Middle East now on bumpers and products that are available.

So while it won't be easy for OPEC to run a refinery cartel, I don't think we can dismiss it because of their control over crude oil.

MS. ELINOR SCHWARTZ (California State Office): To what extent has IEA or any of the other organizations of consumer nations addressed the refinery question?

MR. LUCIAN PUGLIARESI (Director of the Office of Oil Policy, Department of Energy): The IEA is just starting to look into it because there is a problem. They deal mostly with allocation of crude. There is a good deal of leakage in the system in terms of allocation of products.

I'd also like to make a statement of fact. A great deal of the ills the refining industry suffers from now will go away with crude decontrol because the refiners will begin to see the right price differentials in the marketplace. In fact, based upon expectations that we are going to have decontrol, over 700,000 barrels a day in upgraded capacity is either under construction or in the advanced planning stages in the domestic market now. So the refining industry doesn't seem to be quite as concerned about whether or not a tariff is going to come along. They see these price differentials as an opportunity to make some money. They see decontrol coming and their rates of return on these investments will be adequate, and they are proceeding down that road.

MR. SCHULER: The refining industry is not monolithic. Some companies are totally integrated, some are partially integrated, some are not integrated at all. And some of those that are not integrated at all and some that are partially integrated are very concerned about the level of profitability of the refining sector. They are very concerned about the idea of the Department of Energy and the GAO and others that perhaps we should lower the barriers and let product come in—product they know they will have difficulty competing with. And they are not making the investment that is really required.

MR. PUGLIARESI: Is your comment directed to the basic issues of equity which will be discussed this afternoon, or do you think the inefficiencies that will be generated by a tariff are worth it? Because if that is the case, a tariff may be very appropriate. You may have to have a program that has the kind of capacity that will not be affected during the interruption. I think many questions of equity get mixed up with this, and we ought to try to separate them.

MR. SCHULER: I am perfectly willing to leave the equity argument until this afternoon, although any evaluation of national security ought to take into account equity and competitiveness and the whole thing. That is what national security and the economy are about. But I think there is a real national security problem. And I don't use "national security" in the limited sense of talking about an interruption—that is only one facet of national security. All the other facets that we have examined in the study also should be borne in mind. I think the case is clear that this is one of the areas of energy policy where we are not now dependent upon foreign suppliers. To allow ourselves to become dependent upon foreign refiners in the one area where we aren't yet dependent is to me the height of folly. It is exactly the situation that developed with respect to crude oil.

Not so many years ago people were probably sitting at this same dais and talking about whether we should continue our imports of cheap foreign crude oil or develop alternatives. And the argument inevitably came out, or unfortunately came out, "Continue to rely on the cheap foreign crude oil." It didn't stay cheap; it didn't stay secure. And my point is that 10 years from now we will look back and see the same thing happen with respect to refined products.

MR. PUGLIARESI: There is one flaw with respect to that argument: That is, crude oil is the scarce resource. Whether we allowed cheap crude oil imports or developed alternatives then has nothing to do with our ability to develop them now. Refining capacity is not a scarce resource. You can go to a manufacturer tomorrow and he will build you a refinery. Crude oil is a scarce resource. That is the fundamental difference that gets wrapped up in this argument, and that is what I am concerned about when we examine this as a matter of policy.

MR. SCHULER: You are saying you can get a refinery off the shelf. Having been through some refinery construction projects, I can say it isn't quite that simple. There are lag times that could put this country in a very, very serious situation. A consumer goes to the filling station and can't get any unleaded gasoline because the capacity hasn't been built—our ability to project demand for products has been demonstrated time and time and time again to be very, very poor. And I don't think that anybody in Europe or the Caribbean or Canada is going to be able to project our demand any better than we are.

MR. BANTA: I just wanted to comment on Mr. Schuler's last observation about the uncertainty of all this business. I guess I was as much troubled as anyone looking at it from a public policy point of view. Even if one were to accept the argument for some level of protection, the next question is overwhelming. What level? How one draws the line and where and how many of the factors one takes into account and what weight one assigns to each factor become an almost overwhelming question.

I note that while Mr. Schuler feels most strongly about moving in this direction, he is also most skeptical about DOE's drawing these lines. It is an almost overwhelming problem. Nothing said this morning has done anything to relieve my pessimism about coming up with a rational solution.

MR. JOHN DANSBY (Ashland Oil Company): I have a question for Mr. Horwich. In a recent letter to the *Wall Street Journal* you stated that approximately 50% of the cost of decontrol could be passed through. We have seen a number of these estimates over the years, and at the same time the level of protection which perhaps could be measured by the entitlement credit has also flushed away. Could you tell me: Was the 50% estimate based on the situation six months ago?

MR. HORWICH: That was a current estimate.

MR. DANSBY: Of over $4.00 a barrel then?

MR. HORWICH: Yes.

MR. DANSBY: So you are saying that the refiners will have to absorb $2.00 per barrel additional, whereas a year ago it was far less and they were speaking of a much lower figure?

MR. HORWICH: That's correct.

MR. ROBERT KANE (Robert Kane Associates): Following up on the previous question, Mr. Horwich, the 50% credit which you say is something used in a recent estimate—does that track with this 750,000 barrel refining loss, or wouldn't the number be somewhat greater? I can't believe that a 5¢ per gallon margin wouldn't have a greater impact on domestic refineries.

MR. HORWICH: You think the reduction in output would be greater with just a 50% increase?

MR. KANE: I am suggesting it would be, yes.

MR. HORWICH: I am not sure about the exact reduction of refinery capacity. That was obtained from a different model than the one from which our price pass-through estimates were derived.

MR. KANE: What do you feel today's domestic industry is making?

MR. HORWICH: I don't have a ready number on that. To go back to the original question, I believe the 750,000 barrel refinery cutback is compatible with a 50% price passthrough. But I'd have to check it out to be sure. The models used were not the same.

Part II

The Case for Expanding U.S. Refinery Capacity

Presentation by
Senator J. Bennett Johnston

Thank you very much for putting on this conference today on this very interesting and timely subject of refineries.

I say that really a little bit wryly because refineries, as vital as they are for the country, and as direct an effect as they have on Americans and their energy supplies and the economic future of the country, are such an arcane, esoteric subject that it is hard for people to come to grips with the issue.

Indeed, it is very hard for Congress and the Executive to come to grips with. And I do think you do us a real service by trying to focus on this difficult subject.

I have prepared a paper on "Domestic Refining Policy," outlining the problems we face and the prospects for their resolution, which has been distributed to the conference attendees and will be included in the conference record following my remarks.

My sponsorship of the refinery bill is a little bit enigmatic in a way. I am from an oil state, and everyone knows that everyone from an oil state is supposed to be completely for deregulation of all kinds and at all times. I mean, the free market—at least according to our image as oil-state senators—will solve all problems if you just let it function.

Of course, consumers, on the other side, are against inflation and the high cost of petroleum and the high profits. So here I am, swimming upstream as it were, sponsoring a refinery bill which involves an import fee of 3¢ a gallon, which would have a cost to American consumers and which contains other provisions that have the effect of continuing regulation.

How could it be that I am sponsoring such legislation in the year 1980? Well, sometimes I wonder myself.

But the fact of the matter is we must have more refining capacity, not less. We must have more of the right kind of refining capacity which we badly need, and we are not going to get it, in my view, without some form of regulation and without some form of subsidy. And if we want the independent sector of the refining industry to survive, then we must have some kind of protection for them. It is just as simple as that.

Now, the central question behind the whole issue of refining capacity is: Is it worse to be dependent upon a foreign refined product than it is to be dependent upon foreign crude oil?

We all know, of course, that we are dependent to a very large extent on foreign crude oil, $70 billion worth last year, probably $100 billion worth this year. So if you are dependent, what difference does it make? Why not be dependent on refined capacity as well as crude?

Well, the answer to that, I think, is quite clear. It is a multifaceted answer.

First of all, there is the question of jobs and balance of payments—not the most important reason, but a significant reason when you consider that for every million barrels a day of refining capacity which you export, you are also exporting with that 31,000 jobs on the average, according to the Department of Energy. Construction jobs are 12,000 over a period of three years for every million barrels a day of capacity. So the export of that capacity is an important consideration.

Secondly, the U.S. demand for refined product is different than it is anywhere else in the world. Nobody else produces or uses unleaded gasoline. We are the only country in the world that does. So to the extent we are dependent upon a foreign refiner, we would be dependent upon a single source, because we could not then turn to other sources for the unleaded capacity.

We need much more unleaded capacity. Demand for unleaded gasoline will peak in the 1980s at between six and seven million barrels a day. That is more than a 100% increase over what we have right now. That is going to require about 3.5 million barrels a day of increased capacity for unleaded gasoline.

If we were dependent on foreign refineries for most of our refined product, then the Strategic Petroleum Reserve would be rather useless for us or to us, except to the extent that we could ship that out and bring it back in. So if it is important to have it grow—and I think it is vital to us—then we need the refining capacity for that as well.

To the extent we are dependent on foreign refiners, we are dependent on their pricing policies. There is no antitrust law internationally among foreign countries. They are free to be exploitative. They are free to be politically manipulative. They are free to form a cartel and, indeed, in my view would do so to the extent that they could. And OPEC pricing policies are a very good indication of what they can do and have done with respect to cartels and pricing policies.

But most important, in my view, is the need to have domestic refining capacity so that we cannot be subject to a selective embargo. We have had testimony before our Committee on Energy and Natural Resources on this very subject. For example, we considered testimony from G. Henry M. Schuler of Melvin Conant and Associates, who is the author of an excellent study on national security implications of reliance upon foreign refined products. He and other experts who have analyzed the problem point out that it is much easier to focus a politically inspired embargo or boycott on the United States through refined products than it is through crude oil.

There is, as I am sure most of you know, a theory, and I think really a fact, about the crude oil supply in the world. The theory is that the world market for crude oil is like a pool, and it doesn't particularly make any difference who is exporting to whom or where you are buying your oil from so long as there is enough in that pool. The recent boycott of Iranian products is a good example of that. It hasn't reduced supply to this country one whit because when we quit importing oil from Iran, Iran began exporting to someone else. That made the oil available to other suppliers, and the spot price was able to go down on that account. So it doesn't make very much difference with crude oil.

In 1973, there was an embargo on sales to the United States established by the Arab producers. But it wasn't really effective because the major international oil companies were simply able to shift supplies around and there were vast leaks in, if not total avoidance of, the boycott of crude oil sales to the United States.

Now that would not be so with respect to refined products, because to the extent that you are dependent on high quality refined products, you have almost a one-on-one relationship between suppliers and sellers and you can be selectively denied product. So if, for example, we were dependent on Saudi Arabia for unleaded gasoline—if they were refining it—and they wanted to institute an embargo for political reasons, they could do so much more easily than they can now.

These were the findings of the witnesses who testified before us, and I think these findings are eminently correct.

We are about two million barrels a day short in refining capacity now; our need is going to increase, not only for unleaded but for heavy sour crude as well. If you look at the figures, the demand for sweet products is going up; the availability of sweet crude is going down. In 1975, for example, our domestic supply was estimated at only 32% heavy sour. By 1979, four years later, it had almost doubled to 55% heavy sour. And the trend line continues to go up for our domestic supply.

Worldwide, the supply of sweet is going down rapidly. The proportion of heavy sour is going up from a very high point already. For example, it has been estimated that 85% of OPEC's reserves are heavy sour; only 15% are sweet.

In spite of that overwhelming abundance of heavy sour oil in OPEC reserves, 50% of what we import now from OPEC is still sweet. So the United States, with its need for sweet products and its reliance on sweet crude as far as refining capacity is concerned, is using up the world's sweet crude at an alarmingly rapid rate. It is subjecting us to, first of all, higher prices, because the premium on price for sweet crude continues to go up, and the supply continues to go down. Refining capacity must be built somewhere to handle this heavy high-sulfur crude. It will be built somewhere. The question is: Is it going to be built in the United States, or is it going to be built elsewhere? That is the question, because it must be built.

The cost of that is tremendous. The cost of reconfiguring for heavy high-sulfur crude will probably, in some instances, cost more than the original refinery did. It is hard to get a precise proportion because of economies of scale, but in many instances the cost of reconfiguration is higher than the book value of the facilities which they have in place.

By 1985, according to DOE, we will need three times the conversion and upgrading facilities that we do now. So we must ask the question: Who builds them?

Not long ago I was on a refinery panel with the Secretary General of OPEC. He was very plain and very clear about the intentions of the OPEC countries. He said they are going to build or want to build refining capacity in the Middle East and in Africa. He stated that when they do—all of this was cloaked in the nicest, sweetest language of friendship, but the meaning was very clear—they are going to require you to buy their refined products if you want their crude. And when they get you dependent on not only their crude but their refined product, then you really are dependent and your policies are hostage to someone else.

If we don't build that capacity now, we never will build it. So it is going to be built. Somebody has to build it because we need it. And the question is: Are we going to do it and do it now, or do we let OPEC or someone else do it?

Well, what prevents refineries from being built now?

First, of course, there is simply the unreliability of domestic refining policy. I remember the Energy Company of Louisiana (ECOL) refinery in Louisiana, the biggest grassroots refinery that had ever been built up to that point. It was greatly celebrated by federal energy officials and by other energy experts around the country. "Isn't this wonderful," they said, "that we now have a new grassroots independent domestic refinery."

Well, a couple of years after it started they had to sell out their independent refinery to one of the majors because of a change in the entitlements law when the so-called reverse entitlements rule came in. So FEA ran out of business intentionally the very people that they had encouraged. And this unreliability in FEA policy, I think —or the perception of unreliability—requires that we come up with a definite refinery policy and put it in the law so that people can rely upon it.

Perhaps the biggest thing that is going to be lost, that is going to be a disincentive for domestic refining, is the loss of the protection for domestic refiners afforded by the entitlements program. Under the entitlements program, which phases out completely by 1981, the $5.00 a barrel advantage which domestic refiners now have for their average acquisition cost of crude will be gone. About $5.00 a barrel will be lost. And the domestic laws—OSHA, Jones Act, Clean Air Act, minimum wage, and all those other laws which increase regulatory delays and which are costly to comply with—will be left in place of the present $5.00 advantage.

In addition, we are in a time of greatly rising cost. Since 1973 there has been a 57% increase in the cost of construction of refineries, a 200% increase since 1962. So you have a time of rising costs, increased capital commitment, lower profitability—and it is even worse with the small and independent refiner, because he is also losing his small refiner bias. That small refiner bias netted independents $837 million last year. In addition, they got even more than the $837 million through special exceptions and the special hearing process. So about $1 billion went to small refiners, which they are losing.

On top of that, the small refiner will be losing, by 1981, the buy/sell protection.

So smaller refiners are really being phased out unless we do something. They absolutely, in my view, cannot survive. And even the big refiners are losing their $5.00 a barrel advantage that they have over foreign refiners.

So I have introduced S. 1684, which is designed to accommodate these needs and these problems.

Let me hasten to say that S. 1684 is probably not going to pass in precisely the exact terms in which it was introduced. Nevertheless, the main parts of it, I hope, will.

Let me just give you a quick rundown on what it does.

The 3 ¢ a gallon or $1.26 a barrel import fee provides 3 ¢ a gallon protection for domestic refiners. This will go into a fund which will be paid out in two ways:

First of all, a Jones Act differential payment will be made to those who use Jones Act domestic vessels. They will receive from the fund the additional cost that it takes to comply with the Jones Act because, of course, foreign refiners need not use Jones Act domestic vessels. That is the first source of payment from the fund.

Secondly, there will be a transferable license exempting the holder from the import fee, which we will give to the owners or those who will produce new, expanded, or reconfigured refineries to handle the kind of product which national needs require, which is basically heavy or high-sulfur crude.

We will give a rulemaking power to DOE to vary the kind and amount of this advantage in order to respond to what the particular needs are at a particular time. It may be necessary or desirable at some point to give this kind of aid to those who produce unleaded fuel, or in other instances purely for processing heavy or high-sulfur crude oil.

In any event, the fund will be used to give this advantage, this incentive, to those who build new, expanded, or reconfigured refineries.

We will also have a loan guarantee provision to give up to 75% of the construction cost to those who build new, expanded, or reconfigured refineries.

And, finally, and most controversially, we will continue a buy/sell provision which will give independents access to crude.

I don't have to tell you that if you are with a major oil company you probably oppose the access provision; if you are with an independent you say it's absolutely essential. I hope we can work out a provision—and it must be fine tuned and it will be rethought—but what we want to do is allow the independent to survive.

The argument against it is that independents can go on the free market and buy up all of the crude they wish; it is just simply a question of price.

That is not altogether true. If you have a 20,000 barrel a day refinery, you can't very well go on the world market and buy up a whole tanker load of crude. You know it doesn't come in those kinds of increments. And, unfortunately, the difference between being able to buy in big tanker loads, being able to buy in large amounts, frequently means the difference of having to go on the spot market and paying maybe $41 a barrel or going on the other markets and paying maybe $34 a barrel. And if you are an independent, you cannot stay in business very long buying the highest priced spot crude in the world.

As I say, we want to fine tune and change this access rule to give independents the ability to buy crude at reasonable prices. What does "reasonable" mean? Well, we haven't worked that out completely yet. That is still an unresolved question. But we want to give independents the ability to buy at reasonable prices—not to give them an advantage over the majors from whom they would buy, not to discourage the majors by making it at some kind of concessionary price, but by making it a fair price to give them the right to survive.

We are well aware that the previous small refiner bias legislation went way too far, further than it needed to go, and was unfair and economically inefficient. And we don't propose to make that same kind of mistake with respect to this buy/sell legislation. Nevertheless, I think it is essential that we have crude oil access provisions in the bill.

We may or may not put further provisions with respect to oil import quotas in this legislation, depending on what happens with the provision relating to import quotas which I added to the windfall profits tax bill.

In my view, quotas will not work. It is like trying to tell people to line up alphabetically by height. It just can't be done. There is no way you can put on a quota and have it work except in a most inefficient way.

So we may or may not want to deal with that, depending upon what happens with the windfall profits tax bill. There are questions of jurisdiction, and we don't want to infringe upon anybody else's jurisdiction. But quotas ought to be addressed at some appropriate time.

What is the outlook for S. 1684? Well, I hope and I think that this Congress will be enlightened enough to pass the legislation because I think it is essential certainly for the survival of independents, and I think it is essential for us to achieve the kind of new and expanded and reconfigured refining capacity which we must have in this country. The provisions which offset Jones Act costs related to supply to import-dependent areas, frankly, are appealing to people in the Northeast, to congressmen and senators in the Northeast. Indeed, last year we had 32 senators who signed a letter urging a refining policy basically along the lines of this refining legislation.

I think it is possible to pass this bill. I think it is not only desirable but it is absolutely necessary—necessary for jobs, for the balance of payments, for strategic considerations, for our foreign policy independence.

If we are to avoid seeing this great nation further humbled and further submissive and further dependent on foreign sources, particularly of refined products, we've got to make these decisions now. These are not decisions which we can put off. To put off decisions on refining policy for a period of a couple of years would be in effect to make the decision by putting it off. Because putting off the decision means that we will forever lose the capacity, the ability, to build those refineries ourselves. Once you are dependent on foreign sources for the refinery product as well as the crude, you lose that ability to build refineries yourselves.

Tomorrow—or year after next, at least—may be too late. It is time to make those decisions now.

Again, I would like to thank Americans For Energy Independence for organizing this conference on the domestic refining industry. The issues associated with America's capacity to refine crude oil into petroleum fuels are central to the operation of our domestic economy, to the political survival of any President and to the relationship of the United States to the world oil market with all the foreign policy implications that go along with that. Given these relationships, it is incredible that it has been only because of the persistent efforts of several of you here today that the refinery policy issue has been brought to the Administration's attention. I hope you will keep up the good work.

Domestic Refining Policy

By Senator J. Bennett Johnston

Let me review the nature of the problems we face in the effort to insure a sufficient and reliable national supply of petroleum products.

Need for Environmentally Acceptable Fuels

Concern for the quality of our environment has produced particular demands on petroleum fuels which are unique to the United States. We require the lowest sulfur levels of any country in the world for our heating and boiler fuels. Each year our automobiles require a half million barrels per day *more* unleaded gasoline than the year before. In 1979, 40% of the motor gasoline consumed in the United States was lead-free.

We have placed these requirements on the fuels we will accept. It is foolish to expect Caribbean and European refiners, whose local markets do not demand what our environmental laws require, to supply us with products of this quality reliably and at reasonable cost. We must look to certainty of supply ourselves by maintaining a modern refining industry capable of supplying the U.S. market.

Changing Crude Oil Slates

The refinery feedstock available to the United States is shifting towards the higher sulfur, lower gravity crude oils. This shift will continue and accelerate.

In 1975, 68% of U.S. crude oil production was sweet. Today this figure is closer to 45% because of Alaskan North Slope oil and also because the major fraction of new domestic production from now on will tend to be heavy and high in sulfur. Existing U.S. reserves are estimated to be only 42% sweet.

On the world market the situation is even more unbalanced. It is estimated that only 15% of OPEC's reserves are sweet crude oil, yet the largest portion by far of OPEC's production is sweet crude. OPEC producers are already moving to close the gap between sweet and sour reserve to production ratios by requiring purchasers to take sour oils along with sweet and by demanding and obtaining premiums of $6–10 per barrel for sweet oils.

So more desulfurization capacity will be needed. We will also need more sophisticated refineries to refine from increasingly heavy crude oils the light products needed for transportation. Too much of our capacity is used for crude oils yielding 60% light products. With the heavy crude oils which will predominate in the future, this equipment can yield only half as much transportation fuel.

Lowered Demands for Heavy Fuels

Refineries which produce predominantly heavy oils will unnecessarily divert crude oil from the production of the gasoline, diesel fuel and jet fuel we will need. As the normal refining yield of available crude oils is resulting in more heavy fuel oil, the demand for heavy fuel oil is expected to stagnate. These fuels have always sold at or below crude oil prices. Coal conversion and, we hope, a return to sanity with regard to the use of nuclear power, will increasingly displace heavy oil. The Carter Administration's principal energy initiative of the Second Session of the 96th Congress will involve acceleration of coal conversion—the "oil backout bill." The Administration hopes to displace the equivalent of 350,000 to 550,000 barrels a day of oil and gas from the utility sector by 1985.

Foreign Competition

If we do not improve our refineries we will inevitably become dependent on others for our refined product supply. OPEC nations have repeatedly signaled their intent to capture more of the downstream profits derived from their crude oil. If they build export refineries, we will be required to buy their output, whether we want to or not, as a condition of obtaining crude oil. So we could become subject to routine price extortion. Beyond this, an embargo on particular refined product shipments to the United States would be far simpler to enforce and, if effective, far more dangerous to our economy. Crude oil is a far more flexible and readily obtainable commodity on the world market than low-sulfur fuel oil, home heating fuel or unleaded gasoline, as our experience of 1977 clearly shows. Certainly the value of the light refined petroleum products—transportation fuel and home heating oil—in the U.S. market far exceeds the value of the crude oil from which it is made. If we permit foreign refineries to obtain a larger share of the U.S. market than they have now, we will substantially worsen our balance of payments.

U.S. Policy

Federal policy for domestic refineries has been constructed in an ad hoc fashion and has sent conflicting signals to the industry and to consumers. This policy has obviously lacked coherence. No one is in charge, and the state of our refining capacity shows it.

Domestic refiners are generally protected from foreign imports through the entitlements program. This level of protection has been as low as $1.35 per barrel and is now probably close to $5.00 per barrel. Yet residual fuel imports are subsidized and, until recently, domestic refiners seeking to supply the most resid-dependent region of the United States were actually penalized for trying to do so. Cost recovery in gasoline is severely restricted while the construction of new ultra small refineries unable to produce transportation fuels is encouraged. The history of the refining industry is a history of plentiful crude oil controlled on the world market and at home by the major oil companies. Now that control may be slipping away and crude oil is scarce relative to refining capacity.

And forcing the policy process towards decisions is a watershed event—the expiration on October 1, 1981 of the authorities contained in the Emergency Petroleum Allocation Act. On that day gasoline will surely be decontrolled and domestic refinery protection will vanish. There will be no small refiner bias and no authority to allocate crude oil among refiners with far different abilities to obtain crude oil domestically and on the world market.

There are many issues which must be addressed when this authority expires, but one of the most important is: "Where are we going with our domestic refineries?"

What We are Trying to Accomplish

The purpose of the legislation I have introduced—the Domestic Refinery Development and Improvement Act of 1979—is to provide a forum for the legislative resolution of these issues. Fundamentally, we are seeking to provide a predictable policy for domestic refineries which is favorable to the investments we believe we need to make to guarantee the availability to U.S. consumers of environmentally acceptable fuels.

Import Fee

To replace the variable, and currently, probably excessive, level of protection for domestic refiners, we propose a 3 ¢ per gallon ($1.26 per barrel) fee on refined product imports which is indexed with general inflation. We are not wedded to this level of protection. Much of the testimony we have had at hearings on the bill suggests that some level of protection is needed.

Refinery Development Fund

The proceeds of this fee will be deposited in a fund dedicated to providing financial incentives to independent refiners seeking to construct or retrofit refineries in order to produce refined petroleum products essential to national needs. In order to ease the impact of the import fee on the import-dependent East Coast, proceeds of the fee will also be used to defray the added cost of waterborne supply by domestic refiners due to the Jones Act. We have had a great deal of advice as to how the money this fee will generate should be used. We remain open minded on the different kinds of incentives it could provide and on the preconditions for eligibility for these incentives.

Loan Guarantees

In order to enhance the terms under which independent firms can acquire capital for refining investments we provide for a program of federal loan guarantees of up to 75% of the cost of refining investments which have characteristics essential to national needs.

Fast Track

In order to streamline the regulatory processes through which a refinery construction or reconfiguration project must move, Title III of our bill provides for the concept of a "priority refinery project." Drafted at the same time the more general notion of a "priority energy project" was being developed in the Senate, the provisions of this title are similar to legislation establishing an Energy Mobilization Board (EMB) currently in House-Senate conference. I am acting as Chairman of the Senate conferees in that conference. I expect that the outcome of the EMB conference will substantially determine the future of any similar provision in our domestic refinery policy.

Crude Oil Allocation for Small Refiners

Perhaps the most difficult issue to address in this legislation is the issue of crude oil access for smaller firms in the refining business. Of the approximately 17.8 million barrels per day of refining capacity in the United

States (including refineries on Puerto Rico, the Virgin Islands and Guam), 11.8 million b/d is owned by 15 major refiners, 2.4 million b/d is owned by seven large independents, and 3.6 million b/d is spread over 145 small refiners. Of the 64 new grassroots refineries built in the U.S. since 1974, 58 have had capacities under 30,000 b/d, five have had capacities between 30,000 b/d and 40,000 b/d and one had capacity above 40,000 b/d.

It is no secret that federal policy has been very favorable to these very small refineries both in terms of direct financial subsidies and indirect subsidies through access to crude oil. It is also no secret that the most important of these subsidy programs end when the Emergency Petroleum Allocation Act expires in 1981.

Our bill attempts to open the discussion concerning access to crude oil for small refiners in the 1980's. Again we are not insistent on the provisions of the bill as introduced.

Since the development during 1979 of a two-tiered world market pricing structure, we have found that several not-so-small refiners, including Ashland and Union Oil, have severe crude oil access problems. As a result, it appears that the Administration may have to make up its mind with regard to possible revisions in its current regulations governing crude oil allocations.

These are all aspects of the same problem: scarce crude oil and uneven ability to gain access to it. A small, independent firm cannot simply go out and buy a cargo of crude oil. On the other hand the major oil companies are still well positioned to acquire crude oil, despite the changes which have taken place in both the domestic and the world crude oil markets. In the past this has been the rationale for allocation. It may not be sufficient in the future.

Do we need all 145 of these small refineries? Our hearings are not conclusive one way or the other. Our objective should be to emerge from the transition which will take place with the expiration of federal controls with a competitive, modern refining capacity which meets the needs of U.S. consumers, particularly for light, clean fuels. Because of the danger of a heavier than necessary dependence on imported crude oil or imported refined products we cannot afford to subsidize the continued operation of inefficient equipment.

Prospects

I am pleased that we have attendance at this conference of representatives of the Administration. I am also heartened by the private signals I have been receiving from within the Department of Energy. The Department has been of several minds on refining issues and, frankly, this has rendered them irrelevant to the debate on the development of a coherent refining policy. I hope they will soon join us and participate in a constructive way.

I am also hopeful that we can continue the very cooperative relationship which has existed between representatives of regions in which the main refining centers are located and those of petroleum consuming regions. The development of an effective refining policy is very much in the interest of each of these regions. I also hope we can avoid the sort of problem which would develop if we end up with different segments of the refining industry in open warfare with each other over the provisions of the bill. There is no need for this to happen, either.

We will find out to what extent these problems really are problems, because we intend to move the legislation. The underlying issues this bill addresses are very real and demand the attention of policymakers. Even if we wanted to, we could not avoid the effort of devising solutions which will work in the real world.

DISCUSSION

MR. STEPHEN GOLDBERG (Office of Senator Bentsen): Would the transferable license and loan guarantee provision of your bill apply only to the small refiners or to other segments of the industry as well?

SENATOR JOHNSTON: Well, as drafted it would not apply to the majors. The transferable license would be available to refiners to the extent they built the kind of capacity we need; that is, principally heavy high-sulfur capacity. Several witnesses at our hearings have suggested that these benefits should apply to other parts of the industry, specifically to the majors.

MR. KIRBY BRANT (House Subcommittee on Commerce, Consumer and Monetary Affairs): The conventional wisdom is that the plight of the independents goes to the ability of the majors to subsidize their refining operations out of subsidies or profits. Do you think that still pertains and, if so, does it stem from government policy at all?

SENATOR JOHNSTON: The question of to what extent the majors are able to subsidize their refineries through other parts—through transportation or production—of course is an unsettled question.

My own view is that that was clearly true a few years ago, but I think it is probably no longer true. I think the majors require that each one of their operations be a profit center unto itself.

Whether it is true or not, however, it is clear that the independents cannot survive without some protection. Whether you attribute that to the majors being able to out-compete them from production or marketing or whatever, or whether it is simply an access to crude, it adds up to the same thing, and that is that independents will be phased out without some kind of protection.

I think independents are an important element in the marketplace. I also think basic fairness says that they be allowed to survive because we have encouraged them to exist and come into existence by government policy, so it would be very unfair and anticompetitive to phase them out.

Part III

Refinery Policy Economic and
Equity Issues

Presentation by Stephen E. McGregor

I will just provide a brief summary statement on refinery policy economic and equity issues as we are framing them within the Department of Energy. To begin with, I would like to associate myself with Senator Johnston's remarks at lunch. The Department of Energy has not yet arrived at a final position on the major current refinery policy issues, although I do anticipate that some action will be taken in the very near future.

I should underscore our understanding of the necessity to take these actions in light of the phased crude oil decontrol schedule which, as most of you know, results in complete decontrol by September 30, 1981.

Before I get into the refinery issues which are currently under analysis in the Department of Energy, let me state that you should make no mistake that the Department of Energy and the Administration as a whole understand the need for a viable, healthy, competitive domestic refining industry. There is no intention on the part of any agency within the Administration to make any policy decisions that do anything but that.

The context in which we are currently analyzing refinery economic and consumer equity issues takes into account that since 1958 we have had federal policies in effect which have gone beyond the normal workings of the marketplace in terms of fostering and at times discouraging particular types of refining development. I am referring initially to voluntary and mandatory import programs, and then, in the post-1973 embargo era, to the entitlements program, offshoots of the entitlements program and other crude and product pricing and allocation programs which have had serious ramifications with respect to how we find the U.S. refining industry structured today.

The U.S. refining sector does account for 80–90% of the petroleum products consumed within the United States. That should be contrasted with a world refinery configuration of between 60% and 65%.

So on the whole, we have endorsed policies which have fostered a very large refining capability in this country—probably larger than the market normally would have induced.

We are now reviewing four areas concerning refinery policy in which the Secretary of Energy perceives a need for action. They are:

First, institutional issues and options;

Second, the tariff issue with respect to protection of the onshore domestic refining industry;

Third, what should be done with respect to refinery utilization and configuration, putting the tariff issue aside; and

Fourth, as Senator Johnston said, the most controversial issue of access to crude oil by various refineries in the United States.

Under the institutional issues and options area, what we are looking at are DOE's crude and product regulations as they exist now, keeping in mind that we are in a period of phased decontrol. The primary constraint we see in this area is our ability to provide investment incentives for refineries, particularly for unleaded gasoline. We are taking a look at what we should do in terms of further adjusting the small refiner bias provision.

There are other issues that fall into this category, such as environmental constraints with respect to siting, air emissions, and what have you.

Moving into the area of protection for the domestic refining industry, obviously until recently we have had tariffs on imported product that have exceeded the tariff on crude which, when added onto the entitlement credits, have given a further advantage to U.S. refining capacity. When you couple that with the small refiner bias, you will see we have had 400,000 barrels of new capacity constructed in this country over the past six years by basically 40 small refineries. That means 10,000 barrels per refinery. And certainly we have some concern over the economies of scale that might not exist in that small a refinery operation.

The types of tariff levels that we are looking at run from zero to $2.00. Our analysis has indicated that if we had no tariff protection, we would expect 500,000 to 750,000 barrels of domestic refining capacity transferring from onshore U.S. refineries to, most likely, the Caribbean, some of the Mediterrean refineries, and perhaps Canada.

Now, the vast bulk of that capacity, if indeed it does move, is in the residual yield area.

In terms of economic policy issues, no doubt a tariff does impose a cost on the nation in terms of economic efficiency. A $2.00 tariff under our analysis indicates a cost in terms of economic efficiency of about $100 million a year. Even more than that, you get a consumer cost through a $2.00 tariff which runs in the range of $5-6 billion a year, and that is a direct income transfer between U.S. consumers and U.S. refinery operations.

There are of course other equity issues, such as the Jones Act, which Senator Johnston mentioned. I should mention that Jones Act shipment of product in the intercoastal trade of the United States accounts for only 8% of total product consumed, the remaining 92% being transported primarily by pipeline.

And, we have to take into account the trade-offs which Senator Johnston mentioned with regard to balance of payments.

With respect to refinery configuration, it is true that the barrel of crude is becoming heavier and more sour. We will first have to assess whether the unfettered marketplace will allow refineries to make the appropriate investments that will give them the downstream capability to produce the lighter products as well as the ability to refine heavier and more sour crudes.

Alternatives to the free market option would be to investigate subsidies and credits for refineries to induce investments.

Finally, moving into the area of crude oil access, there are basically three types of crude oil access which we see in the international marketplace.

The first concerns those majors who do have preferential access to a particular source of crude. The classic example would be the ARAMCO concession arrangement with the Saudi Arabians, where crude, which is now going at $26 per barrel, is substantially below the marginal contract price for foreign crude and also below spot market prices.

The second area we are looking at concerns those refineries or oil companies in this country who are purchasing the large majority of their crude oil on the spot market. This is the converse of the ARAMCO situation. Union Oil obviously had precisely that problem and was provided temporary relief.

The third international crude oil access issue concerns those refiners whose competitive position is jeopardized by unilateral government actions, such as our decision not to import any more Iranian crude oil. Ashland and Hess fall into this category.

On the domestic side there are also crude access issues, and those fall into the supplier/purchaser relationships.

The current policy options under review again run from the free market option, "let the market bear what it will," to some sort of equity adjustment with respect to those refiners who are placed in a precarious position. And the problem, once you start approaching and trying to solve the equity issues, is that you set up incentives for those who are benefiting from such action to go out and purchase higher-cost crude.

Those are the refinery policy areas we are looking at. I do hope resolution will be forthcoming in the very near future. Obviously, there is a need to do something well in advance of total crude oil decontrol. And I won't preclude the possibility that the forum in which those solutions will occur will be the Congress versus the Administration.

Presentation by William H. Magee

One of the major issues confronting our economy is productivity, or how to get the most out of our available assets. This issue involves both capital formation and the allocation of that capital to competing projects.

Assuming the availability of sufficient capital for investment in the petroleum industry—which is somewhat questionable in light of the windfall profits tax, proposed divestiture, continued price controls, and other such punitive measures—the spotlight then focuses on how and where to apply this capital. Within the refining sector, Atlantic Richfield believes there are three major areas that should be *national* priorities: the ability to process low quality crude, increased production capacity for unleaded gasoline, and upgrading the heavier petroleum products associated with the bottom of the barrel into the more critical "light-end" products such as gasoline and heating oil.

Of these concerns, we think that the third, upgrading product mix, is probably the most self-correcting. Current market forces, if allowed to operate, will encourage this trend. The government has reinforced this trend by issuing the gasoline tilt rule and by legislatively curtailing demand for industrial and commercial fuel oils through the Fuel Use Act and, indirectly, the Natural Gas Policy Act.

Unleaded gasoline production, our second concern, is another matter. The refining industry has been hard pressed to keep pace with the dramatic increase in the demand for unleaded gasoline. A major obstacle, as we have repeatedly asserted, is the federal price controls that do not allow for any return on investment capital except for that which a refiner theoretically earned in May 1973, the base period for all pricing provisions. Inflation, of course, has eaten away at these historical margins. For any given refiner these margins may or may not have been adequate, depending on his particular circumstances seven years ago.

Somewhat to its credit, DOE has recognized these regulatory obstacles and has moved to mitigate them through such regulatory "fixes" as gasoline tilt and the recently implemented pricing incentive of 2¢ per gallon for incremental volumes of unleaded gasoline. Welcome as they are, however, these fixes are at best short-term; although they permit a more equitable allocation of costs, they do not by themselves stimulate the new investment in unleaded capacity that will be needed by the mid-eighties.

This brings us to what Atlantic Richfield believes is the nation's primary investment necessity in the refining sector; increasing the capacity to process lower quality crude oil.

The most fundamental and consequential development for the country's refining industry is the changing quality of available crude oil. Crude oil production, both at home and abroad, is supplying a growing percentage of crudes that are heavier in gravity and higher in sulfur content (more "sour"). This trend will increase over time, with a resulting twofold impact:

First, fewer refiners will be able to handle the lower quality crude oil that will be available, since most domestic refineries are designed to run lighter and sweeter crude oils that yield more gasoline and other low-sulfur products.

Second, heavier crudes produce a higher yield of less critical products, such as residual fuel oil and asphalt. Therefore, it will take more crude oil to produce a given amount of the critical light-end products such as transportation fuels and chemical feedstocks, the uses for which petroleum is best suited.

The problem of processing sour crude is certainly not new. In addition to our own experience, the Administration itself was concerned at the outset of 1976 with the ability of West Coast refineries to handle sour Alaskan crude oil from the North Slope. Although this concern was well founded, nothing came from it. In 1975, imports of sweet crude into the West Coast were 544,000 barrels per day. These imports are still running at about 425,000 barrels daily, while we continue to ship Alaskan crude through the Panama Canal to the Gulf Coast.

The domestic refining industry today is functioning under a giant cloud of uncertainty. Of course, the free enterprise system has always lived with risk and uncertainty. But in recent years government regulation has become a new and all-pervasive variable. The effect of such regulation on the risk associated with capital investments is staggering. Although these regulations were initially intended to be a short-term response to the embargo of 1973, they have expanded in piecemeal fashion into a spiderweb that has the potential to strangle the industry and create serious shortages and economic dislocations for years to come.

It is commonly understood by those who have examined the petroleum industry that the regulations, which were intended to combat the effects of a petroleum shortfall, have themselves become a cause of petroleum shortages. Recently, a federal judge, in handing down his decision in an antitrust suit brought against OPEC, put the blame for last summer's gasoline shortage squarely on federal allocation and price rules.

We in the industry know all too well how the regulated environment of the past eight years has undermined our productive capability. During this period, price controls had to undergo a painful evolution before they even recognized many of the real costs of doing business. It was only two years ago, for example, that the cost of depreciation became eligible for recovery. As of today, the regulations still do not recognize all real costs.

But even more discouraging to refinery investments than inadequate provisions for cost recovery and competitive returns on invested capital is the underlying uncertainty resulting from regulatory changes over time. This encompasses retroactive rulemakings, post hoc interpretations, and the ambiguity and contradiction inherent in regulations that have been repeatedly stretched and twisted to assume a more permanent role in influencing refinery policy.

Contending with changing markets and the problems in forecasting petroleum demand is one thing; it goes with the territory. But having to contend also with moving regulatory targets is something else. Trying to incorporate into our forecasts and strategies shifting patterns in federal and local regulations—both price and environmental—is at best a dubious proposition; political winds are indeed treacherous to gauge.

Such uncertainty only compounds the complexity involved in making investment decisions. For instance, in late 1972 we initiated a modernization and expansion project at our Houston refinery involving $200 million of new facilities. These facilities were designed to allow for the processing of substantial volumes of sour crude oil in place of sweet crude and to expand refinery capacity. This investment was largely justified on the basis of a substantial reduction in raw material costs resulting from the price differential between sweet and sour crudes. Unfortunately, subsequent government regulations took away much of the savings by requiring the economic benefits to be passed on to consumers in the form of lower prices.

To make matters worse, in 1977 the DOE adopted an output adjustment factor which wiped out the productivity improvement obtained by expansion of refinery capacity. Is it any wonder that investment in refinery expansions or improvements has been reduced over the last six years?

Let's look at another example. We approved the construction of a fluid cracking unit at our Philadelphia refinery, based on the issuance of the gasoline tilt rule, which corrected the regulations to recognize the increased operating and raw material costs associated with making gasoline. This project is consistent with U.S. national energy goals in that it reduces the quantity of crude oil required to meet gasoline and distillate demand. The projected cost for this unit is on the order of $100 million. A recent DOE inquiry is considering the abolition of the gasoline cost tilt.

One final example. That same inquiry also asked whether the existing refinery fuel conservation regulations should be updated or eliminated. This is unbelievable. The petroleum industry, including ARCO, has invested hundreds of millions of dollars in new equipment to conserve energy. We have improved our energy efficiency by 21%. It is almost incomprehensible to think that the DOE would even consider removing the incentive on which we based our fuel conservation investments, but such is the case.

Although there is some economic risk associated with these investments, it is the *unpredictable and unquantifiable risk associated with an ever-changing regulatory environment* which discourages investments. Regulations do not minimize market risks; they add to that risk the uncertainty of government intervention. The petroleum industry needs improved investment stability!

The answer, very simply, is decontrol. Short of decontrol, there is no other effective and lasting way to provide for more capability to process lower quality crude oil. The concept of a regulated refinery investment incentive is unworkable, inequitable, and an expansion of the regulatory morass. It makes absolutely no sense for the government to try to provide some incentive that first must offset an already existing disincentive. History need no longer repeat itself in the refining sector. This nation simply cannot afford it.

We have heard on countless occasions that decontrol was necessary to encourage investment and restore competition in the petroleum industry. This was not just the industry providing partisan comment. Such governmental agencies as the Department of Justice, Federal Trade Commission and the Department of Energy itself have all criticized the continued control of the petroleum industry.

Most assuredly, we do not need government incentives to remedy the shortfall in refinery investment. All that is required is removal of current government disincentives. We can do the job—if Congress and the Administration will let us.

Presentation by James A. O'Neill, Jr.

I firmly believe there is a need for significant investment in the U.S. refining industry. When one considers all factors—all the costs and benefits—it becomes clear that our failure to make these investments will put us in an extremely precarious position with regard to our future supply of refined petroleum products.

The important issues to be addressed by this conference are:
- Why is additional investment necessary?
- What kind of investment should it be?
- What kind of governmental assistance is required?

My presentation attempts to address all of these questions.

National Security Requires a Strong Domestic Refining Industry

The United States currently imports approximately two million barrels per day of petroleum products. As crude oil decontrol is phased in, and domestic refiners lose their crude oil cost advantage, the volume of product imports will likely increase significantly. In addition to the volume increasing, the variety of products imported will probably increase. More gasoline and distillate fuel oil will be imported. Whereas now well over half of the product imported is relatively low-cost residual fuel oil, the future will bring a higher proportion of high-cost gasoline and distillate fuel oil. Indeed this trend has already begun. Whereas in 1978, residual fuel comprised 67.4% of all product imports, in 1979 the figure dropped to 59.5%. Over the four weeks ended January 18, 1980, this percentage was 56.5. Imports of other products have increased in absolute as well as relative terms. In 1978, imports of products other than residual fuel averaged 653,000 b/d. In 1979, this figure rose to 740,000 b/d, an increase of 13.3%. In the week ended January 18, 1980, imports of other products were 941,000 b/d. Over the most recent four week period, this figure averaged 893,000.

Production Flexibility

We have, to date, depended upon Caribbean and European refineries for virtually all of our imported products. These refineries have been able to produce the product we primarily require— residual fuel oil. Our dependence upon foreign refineries for large volumes of residual fuel is a result of government programs begun in 1957 which made imported residual fuel more accessible and cheaper than domestic production. The general policy of encouraging imports of residual fuel has since continued through special treatment of that product under the Mandatory Oil Import Program and the Entitlements Program. Foreign plants possess the capability to supply this product of appropriate quality (i.e., sulfur content) as long as they continue to have access to the proper quality of crude. As long as low-sulfur crude oils remain available, these refineries can produce the low-sulfur residual and distillate fuel oils required for consumption on the East Coast of the United States. However, they do not generally have the capability to supply these products from the high-sulfur, heavy crudes which are comprising an increasing portion of world supplies. They possess almost no capacity to produce unleaded gasoline, which will represent a higher and higher portion of gasoline demand in the future. Extensive capital investment would be necessary if these plants were to supply their products from the heavier higher sulfur crude oils. It makes more sense for these investments to be made within the United States rather than abroad.

In addition, if the United States does not make the necessary refinery investments now, not only will we become even more dependent upon European and Caribbean refineries; but we will also invite OPEC to expand their import refining capacity in the Middle East and require us to take finished products along with any crude oil they sell us. Indeed, some of the Middle Eastern nations have already begun processing deals with European refiners. In these deals, the refiner refines crude given to him and returns the finished product to the crude supplier. By entering into this type of agreement, the OPEC nations can begin to penetrate the finished product market without expensive and time-consuming grassroots refinery construction. In view of today's situation in the Middle East, such a dependence would be extremely risky. We must send OPEC a signal that we will not allow ourselves to depend upon them for finished products in addition to crude oil. Only by adopting a comprehensive national refining policy which encourages the necessary investments can we send such a signal.

The present turmoil in the Middle East reinforces the need for this country to have strong refinery capabilities. U.S. refineries can be counted on where foreign refineries cannot—for the production of military jet fuel, special oils and petro-chemicals and other necessary products in times of a national emergency. But additional investments are necessary if we are to possess those capabilities— investments that will be much more likely to take place with a comprehensive refinery program in place.

Security of Supply

A second problem with any dependence on foreign refineries is security of supply at competitive prices. U.S. policy has long supported, even subsidized, foreign refineries, particularly those in the Caribbean. Some people argue that these refineries were built to supply the U.S. market and that therefore our energy policies should ensure their continued viability, even at the expense of domestic refiners. Many people have repeatedly warned the government that it has misplaced its confidence—that in times of shortage there is no guarantee of uninterrupted supply from the Caribbean or any other foreign refineries which might have supplied us during better times. We have also pointed out that any subsidization of imported petroleum products, aimed at lowering prices to ultimate consumers, would only be absorbed by foreign refiners in the form of higher prices at their refineries. The record is replete with evidence supporting our contentions. Whenever product entitlements have been granted to foreign petroleum products, prices have risen almost immediately to absorb most, if not all of the subsidy. With regard to supply, we saw in the first half of last year that foreign refiners' only interest is price. During the shortage, the Caribbean refiners were shipping their production to Europe, where prices were anywhere from $5.00 to $15 higher than in the United States. Even following the promulgation of a DOE regulation granting a $5.00 per barrel (11.9¢ a gallon) subsidy to imported middle distillate, a large portion of distillate from Caribbean refineries continued to flow to Europe. There can no longer be any question that these Caribbean refiners can only be counted on for supply if we are willing to pay them more than anyone else in the world. In the long run, this previously "cheap" foreign product will be much more expensive than that produced by refineries here at home.

Finally, in response to the argument that some have made, namely that "the Caribbean refineries were built to serve and supply the U.S. market and that therefore our energy policies should ensure their continued viability, even at the expense of domestic refiners," we need only look at the trend of changing ownership and nationalization movements in the Caribbean refineries to see that such a "moral responsibility" by the United States (to protect the viability of these refineries) would not seem to be justified any longer. (See "The National Security Implications of Increased Reliance upon the Importation of Refined Products," by G. Henry M. Schuler, pages 137—141.)

Economic Security

When considering our need for additional refinery investment in the United States, we must recognize the economic ramifications of increased dependence upon foreign refineries.

First, whenever we import a barrel of light, finished product instead of crude to be run in a domestic plant, we are exacerbating our balance-of-payments problem. These products, i.e., gasoline and home heating oil, are priced significantly higher than crude oil. For example, in June of 1979, the weighted average world crude oil price was about $18 per barrel, but gasoline and heating oil both sold for over $50 per barrel on the Rotterdam spot market. The assertion that there is little macroeconomic impact of product imports proceeds from the assumption that the only imports will be residual fuel oil and that resid will continue to be priced at a level close to that of crude oil. Both assumptions are subject to challenge. There is no reason to believe that resid will be the only import if the U.S. refining industry is allowed to atrophy or fails to meet the growth in demand; as I mentioned earlier, a trend has already begun toward increased imports of more highly valued products. Nor is there any reason why a cartel of low-sulfur resid producers could not significantly increase the price of the one export which is critical to their economies.

This U.S. balance-of-payments drain will exacerbate the decline of the dollar. The decline of the dollar will reduce the crude oil costs of foreign refiners, thereby helping them to undersell U.S. refiners. This will lead to greater product imports, further increasing the U.S. balance-of-payments deficit and contributing to a new cycle of pressure on the dollar. This compounding effect does not exist if the United States buys crude oil rather than products.

In addition, new refineries provide jobs. Our refinery, although smaller than many, nonetheless employed approximately 235 skilled American workers at the peak of construction at the refinery site. Many more were employed in manufacturing and transporting the materials needed to construct the refinery. Indeed, Mr. A. F. Grospiron, former President of the Oil, Chemical and Atomic Workers International Union, has estimated that:

> [T]he practice of building refineries overseas deprives American workers of oil industry jobs. On the average it requires 12 refinery workers for each thousand barrels of crude processed daily . . . . Additionally, according to a study done by Gulf Canada, each refinery job creates 3.5 jobs in closely allied service and manufacturing sectors, and it is the company's opinion that the same holds true in the United States.

Hence, eliminating imports of refined petroleum products through the expansion of domestic refining capacity by two million barrels per day could result in as many as 100,000 to 135,000 permanent jobs for Americans and many thousands of additional jobs for Americans in expanding and upgrading our existing refineries to remedy this shortage. In effect, by relying on product imports we have exported these jobs, which is totally inconsistent with the national goal of reducing unemployment. In addition to providing jobs, new refineries would provide an expanded tax base for the nation and the states in which they are located. New refineries cost millions, in some cases hundreds of millions, of dollars. These dollars would remain in the United States rather than be exported to build additional, unneeded refinery capacity overseas.

What Refinery Capacity is Needed in the United States?

Any discussion of U.S. refinery capacity requirements necessarily must consider these questions: (1) How much additional distillation capacity is needed, and (2) what kind of capacity should it be? Having answered these two questions one must further ask: Who should build this required capacity and where should it be built?

Is Additional Distillation Capacity Required?

There are those who argue that the United States does not need any additional crude oil processing capability. Their primary reasons for this are that there is currently a worldwide excess of distillation capacity and that growth of world demand is expected to slow down considerably over the next 5–10 years. Another argument is that no new capacity should be built because any import quota program would prevent the availability of foreign crude oil for the plant.

As of January 1, 1980, the United States had *17.9 million* barrels of refinery capacity. This capacity, run at an 88% utilization rate (the maximum achievable over an extended period), could process *15.8 million* barrels of crude oil. Demand over the next five years will average somewhere around 18 million barrels per day (excluding NGL) leaving, in terms of absolute volumes, a shortfall of *2.2 million* barrels if no more capacity is built. Absent the construction of new capacity, imports of petroleum products will have to fill the gap. Thus, if we are to merely cut our product imports in half, we would require 1.3 million barrels per day of additional daily capacity. If demand increases to 19 million barrels per day by 1985, as some experts predict, the shortfall will be 3.2 million barrels per day absent additional refinery construction.

Such shortfalls would be filled by refineries currently running below capacity in Europe and the Caribbean. After a point they might be filled by OPEC refineries in the Middle East. We must decide whether it is in the interest of the United States to increase our dependence on those refineries.

What Kind of Refinery Capacity is Needed in the United States?

It is important to understand that consumption/demand patterns are quite different in the United States than in other parts of the world. We consume as gasoline a far greater proportion of our products than other countries. For example in 1978 we consumed 7.4 million barrels of gasoline per day, which accounted for 39% of all petroleum products consumed. By contrast, in Western Europe gasoline consumption was 2.4 million barrels per day, accounting for only 16.5% of all consumption.

In addition to our somewhat unique appetite for motor gasoline, the United States has, because of EPA regulations, an unusually high requirement for low-sulfur fuel oils, i.e., home heating oils, diesel fuels and industrial fuels. Also, because of our disproportionate gasoline needs, we have badly insufficient capacity to produce residual (industrial) fuel oil.

These three peculiarities of U.S. supply and demand take on even more importance when one considers that the low-sulfur, high gravity crude oils which are best suited to fulfilling this country's needs are rapidly diminishing in proportion to high-sulfur, heavier crudes. As the world crude oil mix changes, many refineries, particularly those foreign refineries upon which we depend for most of our product imports, will be unable to supply us products of the required quality. Thus, if they are to continue to supply us, they will have to make substantial investments in their facilities. Even then, we would never be fully assured of receiving products of proper quality on an uninterrupted basis.

While our domestic refineries are generally better suited to make gasoline and low-sulfur fuel oils from these heavier, higher sulfur crude oils, we too have inadequate capacity in this regard. As unleaded gasoline increases as a proportion of total gasoline demand, additional investments will be required to enable us to produce this high octane unleaded gasoline consumed by many automobiles. Further investment will also be needed for desulfurization capacity to produce low-sulfur fuels.

In summary, the U.S. refining industry must make the investments necessary to enable it to produce high octane unleaded motor gasoline and low-sulfur distillates and residual fuel oils from heavy, high-sulfur crude oils.

Who Should Build this Refinery Capacity?

It is essential that we encourage competition in our domestic refining industry. Competition means lower consumer prices. The best way to enhance competition is through the encouragement of independent companies to build, expand and retrofit refineries.

There have recently been a great number of attacks on the smaller, independent refiners. Much of this criticism is based on the small refiner bias, a program under which certain small refiners were subsidized by as much as $2.00 per barrel. As a result, a number of relatively small and inefficient refineries were built. This may not have been in our country's best interests. However, as crude oil prices are decontrolled and the entitlements program is thereby phased out, so too will the small refiner bias be gradually eliminated. At that time, those refineries which were built solely to take advantage of these subsidies will be forced to make significant investments in order to remain competitive or go out of business.

There is, however, a larger group of companies which, despite being small and/or independent, are nonetheless quite efficient. In fact, there is evidence that independent refiners as a group are even more efficient than the integrated companies. In many cases, prior to 1973, the independent refiner was able to absorb higher crude costs and still undersell its major competitors.

Independent refiners are the historical suppliers of the independent marketers, who have consistently been responsible for keeping downward pressure on retail prices of petroleum products. It is clear that the American public as consumers has a direct interest in the continued existence—and indeed the expansion—of independent refiners as a competitive force in the industry.

It is also clear that independent refiners are at a significant competitive disadvantage vis-a-vis the majors because they do not control a captive supply of crude oil. This disadvantage has two parts—*cost* and *access*. Generally, the independent's crude oil is higher-priced than the major's despite the entitlements program. While the integrated refiner may "charge" its refinery the market price for crude, its actual costs of crude are much lower. The cost of crude for the independent refiner, however, is the full market price. In addition, independent refiners are faced with the problem of access to crude oil. They are, for the most part, dependent upon their major competitors for their supplies of crude.

In summary, independent refiners play an important role in the competitive structure of the U.S. refining industry. Their viability should be protected and this group in particular should be encouraged to make the investments needed in U.S. refining capacity.

Where Should this New Capacity be Built?

The geographical distribution of refinery capacity throughout the United States is presently less than optimal. PAD I (the East Coast) is refinery-deficient, producing only about 23% of its product requirements within the district. Some 28% comes from imports and the remainder from refineries in the Gulf Coast (PAD III), the one area in the United States with surplus capacity. PADs II (Midwest), IV (Rocky Mountains) and V (West Coast) have historically maintained a balance between refinery capacity and petroleum product demand.

The shortage of refinery capacity on the East Coast causes several problems:

High Consumer Costs—Essentially one-half of the East Coast demand is met by shipments of refined products from the Gulf Coast to the East Coast. The cost of moving products by pipelines is about 60¢ per barrel. The cost of movement by water is $1.00–2.00 per barrel. Differences in posted prices between the two areas reflect this advantage for the Gulf Coast consumer over the East Coast consumer.

Storage Needs—Another problem occurs in times of supply interruption or embargoes. In the Arab embargo which began in late 1973 and ended after the first quarter of 1974, the East Coast was one of the hardest hit areas in the nation. This persisted until such time as an allocation system could get working. With refineries present in the area, an inventory cushion is provided. Most refineries have 10 days of crude supply and 20 days of product inventory. This permits time for planning to cope with the shortage. It also contributes to the objective of the Strategic Petroleum Reserve program.

Product Mix—The pattern of product requirements on the East Coast differs quite radically from the remainder of the country. Whereas on a nationwide basis distillate and residual fuel oil represent roughly 28% of demand, on the East Coast these two products account for about 47%.

This means that refineries that are needed on the East Coast would be designed to produce significant volumes of low-sulfur fuel oils. If sites on the East Coast were not obtainable, and the capacity were to be considered for the Gulf Coast, one would be faced with constructing a refinery with a product mix alien to the demand pattern of the area of its location plus a high transportation fee of $1.00–2.00 per barrel to move residual fuel by water (residual fuel cannot be pipelined). Historical import protection, although deemed by some to be reasonable, does not take into account such costs and would force selection of a foreign site over the Gulf in most instances.

Hence it is clear that the primary requirements for additional refinery capacity are on the East Coast. While there is a definite need for increased investment in downstream processing throughout the United States, there is also a case to be made for new distillation capacity on the East Coast. A comprehensive refinery policy should take into account both requirements.

Economics of Domestic Refiners vs. Foreign Refiners

Domestic refiners have historically been faced with higher operating costs than their foreign counterparts. These higher costs have provided much of the deterrent to the expansion and retrofit of U.S. refining capacity. These cost advantages of foreign refiners include the following:

Lower crude oil transportation costs—Because of their access to deep water ports, foreign refiners can take advantage of larger tankers to receive crude oil supplies. Such ships result in a lower per barrel transportation cost than the smaller ships required by most U.S. refiners due to more restrictive waterways. This cost differential can be as much as $1.00 per barrel.

Lower capital costs—Due to more relaxed environmental standards and cheaper labor costs, the cost of refinery construction abroad is generally lower than in the United States. More importantly, most foreign plants have been fully depreciated, resulting in lower operating costs on a per barrel basis.

Subsidization by foreign governments—Many foreign plants either receive significant assistance from their host governments through tax exemptions, financing assistance, etc., or are partially or wholly owned by the government. In many cases this allows them to operate on a break-even basis, or even at a loss, with no required rate of return on investment. This, of course, means lower operating costs.

Lower product transportation costs—One of the most significant cost disadvantages of domestic refiners vis-a-vis their foreign competition is the cost of shipping product in required Jones Act tankers. Jones Act tankers usually cost approximately $1.00 per barrel more than foreign flag tankers used by foreign refiners to ship product to the United States. The purpose of requiring U.S. plants to ship product in Jones Act tankers is to preserve and promote a strong U.S. merchant marine. The costs of such a policy, intended to benefit the United States as a whole, have resulted in increased costs to the refining industry.

Several studies have been made in attempts to quantify these cost disparities. Late last year, Pace Engineering, a widely respected independent consulting firm, in a study commissioned by DOE, concluded that an existing (fully depreciated) U.S. Gulf Coast plant marketing product on the East Coast was disadvantaged by up to $2.14 (in 1978 U.S. dollars) per barrel vs. an existing Caribbean refinery. For new refineries on the Gulf Coast and in the Caribbean, this number was as high as $3.00. Heretofore these cost disadvantages have been partially offset by lower crude oil costs for U.S. refiners, as a result of price controls on domestic crude oil, and by import fees on imported petroleum products. Recently, however, the President waived all fees on imported petroleum products, so that this portion of compensation no longer exists. Now, with the prospect of domestic crude oil decontrol resulting in the end of crude oil cost advantages, domestic refiners will find themselves in a very vulnerable position vis-a-vis

long-term contract with an established crude producer. Since 1973, such contracts have been proven subject to interruption by forces clearly beyond the control of the independent refiner. Therefore, part of the guarantee on which investments can be based must be a government program which provides for a sharing of crude, at fair prices, during supply interruptions.

An independent refiner should have access to such crude, but only if it had made good faith efforts to secure adequate long-term crude supplies.

In the course of debate over a national refinery policy, much will be heard from the international major oil companies. However, it must be understood that these companies are, for the most part, foreign refiners. Of Exxon's total worldwide refining capacity, roughly one-quarter is in the United States. The same is true of Shell. In the cases of Mobil and Texaco, roughly one-third of their total capacity is here. Therefore, we must realize that a program assisting foreign refiners will get greater support from these companies than a program helping domestic refiners.

Costs to Consumers of a Strong U.S. Refining Industry

It is clear that there is a cost associated with the development of a strong, modern American refining industry. Under the proper type of refining policy, these costs would be minimized and equitably spread across geographic areas. The concerns that have been expressed about the costs of a tariff program or investment incentives must be evaluated in comparison to the benefits they would bring about. We believe the benefits wholly justify these costs.

However, if, having considered all the benefits, consumer costs of a tariff (or other aspects of any refinery policy) are still believed to be too high, the costs could be lowered to an acceptable range by properly structuring the program.

DOE's refining study examined only fees of $1.00 or $2.00 that would be (1) imposed immediately, and (2) at the same rate on all products.

But witnesses testifying on S. 1684 suggested a number of ways to decrease consumer costs.

One suggestion was to set fees at a lower rate on resid than on other products. This would recognize our historic acceptance and dependence upon imported resid but would at least prevent our becoming dependent on foreign refiners for other products. DOE's draft refining study recognized that the bulk of any consumer costs would result from a high fee on resid. This is true since resid comprises roughly 60% of all product imports and the price of domestic resid is closely related to the price of imported resid.

A higher fee on products other than resid would encourage investments at home to enable us to meet future demands for products which can only be produced by sophisticated refineries and would prevent our becoming dependent on foreign refiners for these products. The consumer effects would be very small, however, since, as DOE's draft analysis states, the amount of imports of these products is small and the price at which these products are sold is set by competition among domestic refiners rather than by the price of imports.

Witnesses on S. 1684 also suggested that fees be phased in, for example, on a fixed schedule of three to five years, as new or retrofitted refineries come on stream, or that there be a permanent fee-free window for an amount of imports equal to some fraction of historic imports. Either of these steps would reduce consumer costs, while providing the degree of certainty needed to encourage investment in domestic refineries.

Another suggestion was to refund any revenue from fees to states which consumed the imported products. The consuming states could use the revenues for any purpose; for example, to assist low-income persons in meeting rising energy costs.

Finally, I would note that S. 1684 authorizes full entitlements on resid and No. 2 heating oil. Such entitlements, already partially in place for resid, could ease the impact of import fees on consumers in areas heavily dependent upon resid or heating oil imports as long as the entitlements program exists.

Need for Legislation this Year

Companies and, as importantly, financial institutions, are extremely wary of the uncertainty caused by unpredictable federal regulation of the refining industry. Senator Johnston, in his remarks when introducing S. 1684, cited a list of federal "Signals to the Domestic Refining Industry Regarding the Climate for Further Investment." The list contains 27 items, which I would like to submit for the record of these proceedings—all negative—and even so, is not exhaustive. Few people are willing to spend the many millions of dollars to build a refinery which can be made non-viable overnight as a result of changes in government regulations. The only major grassroots refinery project in the recent past, the 200,000 barrel per day ECOL refinery in Louisiana (of which Ingram Corporation was a 50% owner), was built to fulfill the country's need for additional refinery capacity to convert high-sulfur, low gravity crude oil into low-sulfur fuel oils. The ECOL project represented an investment of several hundred million dollars and more than three years of effort by thousands of people. With the stroke of a pen the Federal Energy Administration rendered the project uncompetitive through the East Coast residual fuel entitlements program. The plant was subsequently bought by Marathon Oil Company, which has recently spent several hundred million dollars *more* to convert the plant to emphasize gasoline production. As a result, a new entrant to the refining industry was eliminated and, further, there will be no significant addition to U.S. residual fuel production (of which we import over a million barrels per day) from the refinery. ECOL remains a constant reminder of the threat of arbitrary and capricious regulatory actions to any refinery investment in the United States. Only by adopting comprehensive refinery *legislation* can this be mitigated. S. 1684 is absolutely essential in this regard.

Under the President's decontrol program, domestic crude will be rapidly decontrolled between now and October 1981. As decontrol occurs, it will eliminate domestic refiners' crude cost advantage which has offset other cost disadvantages imposed by U.S. government policies on domestic refiners. Foreign refiners will, therefore, begin to increase their share of the U.S. market.

This import penetration will begin this year—taking the form initially of a reduction in the profitability of many domestic refiners to a point so low that new investment in their facilities is impractical and their very existence is threatened. To prevent this undermining of domestic refiners from occurring, we need legislation that will offset the effect of decontrol *as decontrol takes place,* not just after all controls are removed.

If independents are to expand, they must be able to show financial institutions that they will be able to obtain crude oil to process in their refineries. In view of the experience since 1973, long-term contracts with crude suppliers are no longer a credible guarantee that an independent will actually obtain crude oil. Therefore, the guarantee provided by a government program such as the buy/sell is essential for an independent who must borrow capital for refinery expansion or retrofitting.

The fact that the legal authority for the buy/sell program, or the EPAA, expires in October 1981—before any significant expansion or retrofitting will be completed—means that this guarantee is not available. Therefore, independents are stymied in undertaking substantial new investments until statutory authority is enacted to continue a program such as the buy/sell.

Signals to the Domestic Refining Industry Regarding the Climate for Further Investment*

Date, Event, and Documentation

December 5, 1974: Federal Energy Administration petroleum product price regulations deny the inclusions of capital investments by refineries when computing the expenses involved in producing refined products; 39 Fed. Reg. 42368 (December 5, 1974).

July 21, 1976: The FEA proposes rules increasing protection for the domestic refinery because the present level of protection is too low. However, it refuses to raise protection to the level indicated by "security consideration alone." The proposed level of protection is "minimum" according to the FEA rulemaking; 41 Fed. Reg. 30058 (July 21, 1976).

February 16, 1977: Federal Government adopts regulations subsidizing imports of more heating oil to the East Coast. Actions involved cost-sharing by domestic refiners across the nation; 42 Fed. Reg. 9379 (February 16, 1977) and *New York Times* (February 10, 1977).

May 10–12, 1977: Carter Administration statements concerning proposed COET estimate that domestic refiners will be forced to absorb one-third of the cost of COET due to the pressure of world markets on U.S. prices; *see* White House statement: May 10, 1977, *Overall Economic and Budgetary Impact of the National Energy Plan,* as reported in the *Daily Executive Reporter, Bureau of National Affairs,* No. 93, May 12, 1977.

June 10, 1977: Joint Committee on Taxation staff report outlines Carter Administration's estimate that refiners would absorb one-third of the proposed increase in costs due to COET; *see* Joint Committee on Taxation report on *Crude Oil Equalization Tax and Rebate.* June 10, 1977 (House Ways and Means Committee Print).

June 17, 1977: Deputy Secretary of Energy, John O'Leary, states before the Subcommittee on Antitrust and Monopoly of the Senate Judiciary Committee that: (a) the present fee level on imported refined products is sufficient to maintain the competitive viability of the domestic refining industry; (b) if a differential in the $1.88 to $3.26 range, per PACE study, is needed, it would raise a serious question as to whether or not we want the onshore additional refining capacity; and (c) if everything works under the NEP we will not need substantial additions to refinery capacities but that if everything doesn't work, we will have a serious problem in this country; testimony before Subcommittee on Antitrust and Monopoly of the Senate Judiciary Committee.

July 13, 1977: Alfred F. Dougherty, Jr., Director of the Federal Trade Commission Bureau of Competition, in a letter to Senator Edward Kennedy (D-Mass.) states that the NEP may cause refined product imports to rise which will deter "domestic *de novo* refining entry and [create] a relative advantage to foreign refineries." He goes on to say that "the market pressure of these foreign imports could severely depress domestic refinery margins; letter to Senator Kennedy from Mr. Dougherty.

July 13, 1977: House Ways and Means Committee Report states that COET could cause the domestic refining industry to be placed at a competitive disadvantage in relation to foreign refiners, and that the Administration should take appropriate action to protect against this eventuality; House Ways and Means Committee, Energy Tax Act of 1977. H.R. Doc. No. 94-596, 95th Congress, 1st Sess. 79 (1977).

August 26, 1977: Federal Energy Administration issues regulations, effective September 1, 1977, which prevent firms which use the net cost method to compute marine transportation costs in crude from recovering the cost of capital for an equity status vessel; 42 Fed. Reg. 43054 (August 26, 1977).

October 5, 1977: The Federal Energy Administration amends the Oil Import Program and expands the fee-exempt license program for residual fuel oil imports; 42 Fed. Reg. 54255.

February 22, 1977: Al Alm, Assistant Secretary of the Department of Energy, states that the Administration opposes the Haskell Amendment which would clarify the President's discretionary authority to take actions he deems necessary to ensure that imports of refined products do not threaten or impair our national security; letter from Al Alm to Mr. Clyde A. Wheeler, Jr., V.P. of Sun Co.

April 1978: Administration officials indicate that they are seriously considering imposing import fees on crude oil. Crude fees without corresponding fees on imported products would escalate domestic refining cost vis-a-vis foreign refineries; Finance Committee consideration of Dole Amendment, April 10, 1978.

June 1978: Congressional Budget Office report states that crude fees would affect "the scale and competitiveness" of domestic refineries and "might jeopardize U.S. refineries position vis-a-vis refineries in Europe and the Caribbean"; CBO publication (#95-110); Direct Federal Action on Oil Imports: An Analysis of Import Fees and Quotas.

June 4, 1979: ERA analysis memorandum concerning residual fuel oil amendments to Entitlements Program indicates that foreign refineries may receive benefits while domestic refiners are hurt; *see* December 4, 1978 memo from Akin & Gump.

June 15, 1978: DOE officially publishes its proposals to increase the entitlements subsidy for imported residual fuel oil from 50% entitlement to a 100% entitlement. All domestic refiners would be required to pay for this subsidy; 43 Fed. Reg. 26551 (June 20, 1978).

October 7, 1978: Senator Edward Brooke (R-Mass.) states on Senate floor that DOE has promised to propose a suspension of the 63-cent import license fee which will reduce protection for domestic refiners; *see* December 4, 1978 memo.

October 10, 1978: OPEC claims industrialized nations are obstructing OPEC countries from participating in the making of finished petroleum products; *Washington Post.*

October 10, 1978: It is reported that while U.S. refiners puzzle over whether they should build plants at home, OPEC is campaigning for a bigger stake in the world's refining and petrochemical industry; *Wall Street Journal.*

October 12, 1978: OPEC demands aid from oil consuming countries in developing petroleum refining industries; *see Washington Post.*

November 15, 1978: DOE Secretary Schlesinger states his willingness to allow the United States to become more dependent upon imports of petroleum products; Secretary Schlesinger's press conference.

December 11, 1978: At hearing before Senate Energy and Natural Resources Committee, DOE endorses further reliance on additional foreign product imports to deal with sudden gasoline shortages. Senator Jackson criticizes proposal to decontrol gasoline; he believes that the Senate would veto gasoline decontrol; O'Leary's statement and *Washington Post,* December 18, 1978.

December 19, 1978: DOE increases the quantity of residual fuel which may be imported into PAD District I not subject to license fees; 43 Fed. Reg. 59458.

January 2, 1979: Department of Interior releases letter opposing Corps of Engineers approval of Hampton Roads Energy Co.'s plan to build a refinery in Virginia; Copy of letter, press release and *Washington Post* article, January 3, 1979.

January 16, 1979: EPA refuses to grant the Pittston Co. a water pollution discharge permit for the Company's planned Maine refinery. EPA cites the possible danger to the bald eagle as primary objection to the refinery proposal; EPA Notice of Determination to Deny NPDES Water Discharge Permit No. ME0022420 and *Washington Post,* January 17, 1979.

January 16, 1979: DOE adopts a standby mandatory crude oil allocation and refinery yield control program. This gives the ERA administrator the power to implement the program in a variety of ways to affect all or selected refiners. The knowledge that this program could be applied at any time plus the fact that ERA will review the regulations continuously so changes can be made as needed injects tremendous uncertainties into the refinery investment decision process; 44 Fed. Reg. 3418 (January 16, 1979).

January 18, 1979: DOE adopts standby product allocation and price regulations and imposes allocations fractions program. The ERA administrator may order any or all of the provisions of these special rules into effect on a national or regional basis, at any level of distribution, and upon any or all refined products; 44 Fed. Reg. 3928 (January 18, 1979).

March 1, 1979: DOE adopts amendments to allocate increased costs to gasoline on a greater than *pro rata* volumetric basis. However, gasoline controls continue; 44 Fed. Reg. 15600 (March 14, 1979).

*Document inserted in the *Congressional Record* (August 3, 1979, pp. S11671-2) by Senator J. Bennett Johnston.

Presentation by Clement B. Malin

Arguments supporting the need for a U.S. domestic refining policy traditionally are based on one or both of two propositions:

First, additional domestic refining capacity is needed to reduce dependence on foreign sources; and

Second, protection for domestic refiners is needed to assure their survival as U.S. prices rise to world levels.

It is not altogether clear whether traditional arguments, arguments of the 1970s, are relevant for the 1980s. Are we, for instance, more secure importing crude oil from OPEC countries than we would be importing product made in refineries outside the United States but still from the same OPEC oil? What would be the cost of replacing the capacity?

With respect to protection, given the expectation of a chronically tight international oil market, will not U.S. refiners be able to obtain market clearing prices for product manufactured in the United States? And even if we are concerned about future overseas competition, assuming foreign refiners can meet U.S. product specifications, do we need anything more than a stand-by tariff protection system? And then only to offset real cost disadvantages reflecting mandated environmental regulations?

If the traditional arguments used to justify a national refining policy are no longer appropriate, what then should we consider in examining the issue of a refinery policy for the 1980s?

We start with the assumption that it is in the national interest to have a strong and viable petroleum refining industry and to refine in the United States the major portion, but not necessarily all, of petroleum product demand.

In assessing our current refining capability to meet a significant portion of product demand, it is necessary, however, to move far enough out into the 1980s, perhaps to the end of the decade, to be able to determine both what capacity may be needed to supply demand and, within the overall capacity, what capability to produce which products. Only after those issues have been examined from a long-term perspective can we determine whether current refinery capacity and capability are adequate and, if not, how it is we arrive at the desired capacity and capability.

Demand for petroleum products in the United States is changing. Overall demand may well have peaked. Indeed, a reasonable projection for the 1980s would be static growth, or another projection might suggest modest growth until the mid-point of the decade with a fall-off in the later years.

Assuming that current demand of about 18.5 million barrels per day is a reasonable estimate for the next year or two, the adequacy of refinery capacity would not seem to be at issue. The recently completed National Petroleum Council study suggests that by 1982 refining capacity will exceed 19 million barrels per day, which, when coupled with the reasonably assured volumes of residual fuels available for import from the Caribbean, suggests that we have adequate total capacity.

This does not mean, of course, that additional crude distillation capacity will not be built and indeed is not needed. Larger units to replace smaller and less efficient units may well be constructed and even some new grassroots facilities may be built, which will add to the total amount of capacity available. What is being suggested here is that no particular program is needed to encourage a massive increase of crude distillation capacity.

If total capacity is adequate, however, what about the ability of that capacity to handle the changing quality of the available crude oil feedstock and to produce the product yield needed by the changing market?

Reserves of heavy crude oil and reserves of sour crude oil in the United States and in the world in general are greater than reserves of light and sweet oils. OPEC countries have, over the past years, initiated policies to increase the liftings of the heavier and sour crudes to more nearly equate to reserves. In the United States, crude oil reserves are tending towards the heavier and sour ranges, in large measure due to the Alaskan North Slope reserve additions. In the 1980s, this trend will mean for U.S. refiners that they must be able to handle an increasingly higher proportion of heavy and sour crudes. Even if the product yield desired were not to change, therefore, refinery investments would be required simply to permit flexibility in crude oil feedstock.

A good number of refiners have anticipated this need and have already invested significant sums in desulfurization equipment. Other refiners will have to do so or be willing to pay the higher prices for sweet and lighter crudes if they are available to run in their refineries.

With respect to the heavier crudes, additional conversion capacity will be required to upgrade the simple yields of the heavy crude oils, which would favor residual fuel oil, to those products which are required in the current mix, and increasingly in the future mix, of U.S. product demand, gasoline and distillate.

On the demand side, what is projected is simply aggravating the deterioration on the supply side. Projected demand for heavy fuel or residual oil, for instance, is being reduced, both by the simple economics of higher prices for oil generally and by mandated switching under the Fuel Use Act and coal conversion regulations. The switching of 50% of current public utility demand away from oil will have a significant impact on the country's need for residual fuel.

Distillate fuel oils, automotive diesel and jet fuel, are one of the few categories of demand that we can expect to show some increase over the decade, along with petrochemical feedstocks. Conversion facilities, such as cracking or coking, to upgrade unneeded residual oil will be required to provide the desired petrochemical feedstocks and the distillate range of product demand.

Finally, with respect to gasoline, while we can expect a decrease in demand over the decade for gasoline reflecting both price and more efficient automobiles, the quality of gasoline demanded will be changing from its current mix of leaded and unleaded to a high proportion of unleaded by the end of the decade. Additional facilities will be needed here as well, since present capabilities of all U.S. refiners are not sufficient to manufacture the ratios of unleaded gasoline that will be required.

In a free market situation, it seems reasonable to assume that investments to provide the necessary capability within the existing capacity of our refining industry will be made in a timely fashion so that demand will be satisfied. One can make the case, therefore, that the best refining policy would be none at all, simply freeing the market and encouraging industry to get on with the job.

Given the uncertainty, however, of our national energy policy initiatives and guidelines over the past four or five years, it may be appropriate to consider incentives which will signal to the refining sector that we are serious as a nation about developing and preserving a strong and viable refining industry. Such incentives need not be tailored to meet the refining sector alone. Indeed, one can argue that within U.S. industry in general, fiscal incentives, such as tax credits and accelerated depreciation, would enable private enterprise to make significant modernization and expansion investments that would make us once again highly competitive versus our international trading partners.

The current gasoline price regulations provide an absolute disincentive to invest in refinery operations. No provision is made under existing DOE price regulations for recovery of a return on investment.

We might address one final issue which hangs over the U.S. petroleum industry in general, but particularly the refining sector. Given the uncertainty in the international crude oil market, can we be certain that adequate crude oil will be available to import and refine to meet unconstrained demand?

In making projections of demand, we generally do not assume a constraint on supply. If, however, we assume, and it seems to be a reasonable assumption, that OPEC will continue to manage both supply and price to maintain upward pressure on price, it may have some further impact on an already static demand assumption. It is perhaps best to characterize the 1980s as a period of chronic supply shortage, which suggests that from time to time we will face disruptions in the smooth flow of supply. Certainly we will continue to face a tight market with upward pressure on price.

If, in fact, our demand projections are too high because of supply inadequacy, there is at least the possibility that we could have in the United States significant quantities of surplus refinery capacity. If this is the case, it is important that we approach the problem of surplus capacity with our overall energy objectives clearly in mind. We should not, for instance, except perhaps in a brief emergency situation, simply reduce all refineries' runs by equal percentage amounts.

In addressing the issues involving the redirection of crude away from its rightful owners, consideration must be given to the ability of individual refiners to run the types of crude that are available, and also to produce the maximum amount of what will then be priority product demands. This may require some refining capacity to drop out of the system, and it will require the greatest restraint on the part of government and industry to insure that market forces can work to the nation's best advantage.

It is also critical that we distinguish between supply problems and price problems. Responses should be tailored to meet those distinctions.

There is an opportunity for the U.S. Government, through administrative and legislative action, to encourage and maintain a strong and viable domestic refining sector immediately and at minimum cost. A positive approach comprised of fiscal incentives and removal of regulatory constraints is appropriate for the nation's needs in the 1980s. Such a policy will encourage investments which are needed to provide refining feedstock flexibility, meet the changing needs of the marketplace for product, and preserve a competitive and modern refining industry in the United States.

foreign refiners. In this situation it is absolutely essential that there be an import fee imposed such as that in S. 1684. We think that data in the Pace study would support an even higher fee.

Guaranteed Access to Crude Oil at Competitive Prices is Essential to Allow Independent Refiners to Survive or Expand

Lessons from 1973–1980 are that crude oil will be available at competitive prices most of the time, *but* during supply interruptions crude oil will *not* be available to independents at competitive prices. This has been the case since the Iranian revolution. This is true even for refineries such as ours which at the outset had long-term contracts for an amount of crude well in excess of our needs.

Under DOE's existing buy/sell or exception relief programs, independents can buy crude from the international majors at the average cost of the crude imported. Even this crude is unfavorably priced for independents, due to the fact that the entitlements program, designed for the sole purpose of equalizing crude oil costs between refiners with access to price-controlled domestic oil and those with access only to uncontrolled oil, has failed to fulfill its goal. As a result, the average cost of imported crude oil is generally higher than domestic crude of comparable quality. To be eligible for these programs, the independents must make several showings, including a good faith effort to locate crude and a crude shortage exceeding 25% of the refiner's capacity. These programs are authorized by provisions of the EPAA which expire in October 1981.

Without DOE's buy/sell or exceptions programs, a number of refiners of substantial size might have been in serious financial trouble.

As the 1980s progress, access to crude oil will continue to be tight, particularly for independent refiners. Domestic crude will constitute a lower percentage of their total crude runs. OPEC will continue to constrain production of its crude. International majors will continue to reduce third-party sales. The multi-tiered price of foreign crude—highlighted by Union Oil's application last summer for exception relief—may well continue, probable to the competitive detriments of independent refiners including those with a long-term contractual relationship with one or more producing countries.

A guarantee of crude oil supply has always been essential for independents to secure financing for expansions or upgrading of refining capacity. Before 1973, such guarantees could have taken the form of a

Presentation by Frank Collins

Americans For Energy Independence is to be congratulated for sponsoring this meeting and assembling this wide diversity of speakers on U.S. refinery policy. The Oil, Chemical and Atomic Workers International Union joins everyone here in shared concern about the critical dependence of the United States on imported oil. This dependence is compounded by imports of oil products. Oil product imports exacerbate existing problems of U.S. security, employment and the drain of foreign exchange.

Even with today's declining levels of oil consumption, the United States still imports almost two million barrels of oil products daily. This means that about 25% of our total oil imports arrive in the form of oil products refined abroad.

Historically, even a higher percentage of the imported oil arrived in the form of oil products. Prior to 1974, oil products made up 60% of the total oil imported.

The reason for the drop in the percentage of oil product imports beginning in 1974 was the crude oil entitlements program under the Emergency Petroleum Allocation Act. This program offers a subsidy—the imported crude oil entitlements benefit. With the entitlements benefit, it is more profitable to import crude oil and refine it here than to refine it abroad, import the products and lose the entitlements benefit. According to DOE figures, this benefit was $4.39 per barrel as of November 1979.

In effect, the entitlements benefit for imported crude oil is a *subsidy* involving a transfer payment under the entitlements program by the domestic oil producers to oil importers. The entitlements benefit subsidy, foreign tax credits, lower production costs and the recent elimination of the U.S. excise tax on imported crude all add up to a powerful economic inducement to the multinational oil companies to produce oil abroad and import it as against their producing the oil here. It is hardly surprising that, given such a strong set of incentives, U.S. oil imports have grown so much since the Emergency Petroleum Allocation Act of 1974.

The disappearance of the crude oil entitlements program in October 1981, as presently scheduled, would restore the pre-1974 situation when the prices of foreign and domestic crudes were nominally equal. I say *nominally* because U.S. credits for foreign income taxes and other factors will remain to reduce the real cost of foreign oil to multinational companies below that of domestic oil in terms of ultimate profitability.

The choice by the multinational oil companies between refining their foreign oil abroad and importing products as against refining the oil in the United States will be determined by comparative costs. The disparity in effective refining costs comes about by fewer environmental regulations, lower wages and, again, by foreign income tax credits. If the pre-1974 percentage of oil products in total U.S. oil imports were restored, it would mean a growth of oil product imports from the present two million barrels per day to almost five million barrels per day. Increases of oil product imports of this magnitude are possible because of surplus refining capacity in the Caribbean and Europe.

If U.S. refineries are now operating somewhat below 85% of full capacity, the additional oil product imports would mean a drop to around 70% of capacity. For the multinational oil companies which have refining capacity abroad, this would simply mean an interchange of the utilization of foreign refineries, because of their greater profitability, for the utilization of domestic refineries. For purely domestic refiners, particularly smaller ones, the loss of refinery throughput would have catastrophic economic consequences. The effects on domestic refinery employment would be severe. OCAW has 60,000 members employed in the domestic refining industry, giving us a special reason for concern.

For a variety of reasons, not necessarily only economic, the geographical distribution of U.S. refineries does not match the geographical distribution of oil consumption. The Northeastern states have insufficient refining capacity to meet local needs, particularly for residual oil for industrial uses. This region is therefore heavily dependent on oil product imports, a situation that will continue until new refineries are built to serve regional needs.

In addition to the inadequacy of the regional distribution of oil refineries, the refining capacity in place requires extensive modernization to handle a new mix of feedstocks and a new slate of products. Increasing amounts of cheaper heavy high-sulfur crudes are becoming available compared to the higher cost light low-sulfur crudes. Heavy high-sulfur crudes require more complex refineries, but the total existing capacity of such refineries is limited. This accounts for absurdities such as West Coast imports of almost 400,000 barrels per day of light, low-sulfur Indonesian crudes when the region is swamped with its own and with Alaskan heavy high-sulfur crude oil.

Furthermore, domestic refineries are unable to meet the potential demand for high-octane unleaded gasoline, a growing demand which was quite predictable from the phased environmental regulations for new cars. The average octane rating of unleaded gasoline now being sold is barely sufficient to keep new cars from knocking badly. The industry would be in even worse condition if the total demand for gasoline had not declined 8% from last year's demand at this time.

Finally, compliance with air quality standards will require increasing volumes of low-sulfur heating oil. The National Petroleum Council's interim report on refinery flexibility states that there is a low probability for the needed capacity for refining larger quantities of low-sulfur heating oil being installed, based on refiners' assessments of future economic conditions.

I would suggest that we are dealing with two separate questions here and that they have two separate solutions, not a single answer. The first question is how to keep out the flood of imported oil products that will certainly occur if domestic crude oil is completely deregulated in October 1981 as the end result of the phased deregulation plan now being put into effect by the Carter Administration. The second question is building and modernization of refineries. That is, we need to build additional refineries in those areas where refining capacity is critically short, as in New England, and we need to modernize existing refineries to handle demand for the new slate of products needed for the future.

Returning to the first question, OCAW has steadfastly opposed the decontrol of domestic crude oil and the consequent equalization of the nominal prices of domestic and foreign crudes which would create a flood of oil product imports, as I pointed out earlier. We oppose crude oil decontrol for several reasons. They add up to the issue of *economics and equity*, the title of this panel.

The scale of the increases in profits from domestic oil production under decontrol is not fully appreciated. A few figures will show what I mean. The latest data in hand for the controlled prices of crude oil are for August 1979. The prices were $6.09 for "old" oil, $13.38 for "new" oil and $26.01 for decontrolled stripper oil. Stripper oil represents the probable price that all oil would have had under instantaneous decontrol last August. The price increase for old oil would have amounted to $19.92, or 327%. On new oil, it would have been $12.62 or 94%.

The increase of average profits would have been even more sensational. The nationwide average cost of production (including exploration and development) reported to the SEC by the 17 largest oil companies for 1978 was $1.85 per barrel. The companies complain that the methods of accounting understate costs by 100%. If we accept this argument and also allow a 20% inflation in costs between 1978 and 1979, we arrive at a cost of $4.20 per barrel.

For old oil, average profits in August 1979 amounted to $1.89 per barrel. After the hypothetical decontrol in August 1979, the new profits would have been $21.19, an increase of 1021%. The increase in profits for new oil would have been 138%.

As both old and new oil are already in production, these price increases amount to outright gifts to the domestic producers in the pious hope that the proceeds (less "windfall profits" taxes) will be used for exploration and development. At best, this represents a change in business philosophy because the funding for new exploration and development would be taken directly from oil consumers instead of being raised in the capital funds market as is traditional.

The Carter Administration's phased decontrol extends the decontrol process over 27 months, but the results are much the same after correcting for inflation. As we see in Congress at this moment, profits of this magnitude are not easily taxed away.

Incidentally, the proposed "windfall profits tax" is actually a per barrel excise tax on domestic oil. A true windfall profits tax would be on the *profits* of production of all oil, both foreign and domestic, instead of discriminating against domestic oil and providing a new incentive for the production and importation of foreign oil.

OCAW advocates the stopping of the Administration's crude oil decontrol process and the rollback of prices in order to balance equity for consumers against the need for reasonable incentives for increased domestic oil production. The OCAW program would prevent the restoration of the pre-1974 situation and the bringing about of a large increase in imported oil products into the United States.

Now to return to the second question of refinery improvement and development. As we have seen, the domestic refinery industry as a whole is not in healthy condition. It lacks capacity to supply local needs for products in certain areas of the country, notably New England. It is unable to handle today's mix of crude oil imports and to produce the needed slate of products for the future.

The response of the domestic refining industry to these questions has been to call for an import fee on foreign refined oil products. At one stroke this would discourage the import of oil products and it would increase the profitability of the domestic refining industry so that needed modernization would be encouraged.

Senator Johnston has introduced a bill—The Domestic Refinery Improvement Act—that would provide an import fee level of $1.26 per barrel or 3¢ per gallon. It is estimated that the $1.26 fee is roughly equal to the average difference in the profitability of domestic and foreign refining. The revenues from the fees would be set aside as a fund to provide incentives for building and modernization of domestic refineries.

The total cost to consumers would not be just the amount received by the U.S. Treasury, but the cost of the corresponding rise in the price of domestic oil products in the protected market. The take of the U.S. Treasury would be $1.09 billion per year based on average product imports over the period 1975–1979. The additional amount received by domestic refiners would be $7.56 billion per year assuming that the 1977–1979 average production continues. The total cost to consumers would be $8.65 billion per year or $152 for a family of four.

The domestic refining industry has become a brilliant performer as a profit-maker, as was brought out by Congressman Rosenthal last week. His figures showed an increase of more than 800% in refiners' profits on home heating oil between September 1978 and September 1979 (the last date for which DOE data are available). It is likely that these increases in refiners' profits on home heating oil cited by Mr. Rosenthal are matched by large increases in profits on other oil products.

Under these circumstances, it is an incredible argument that an oil products import fee is needed to enable refiners to finance the needed modernization of the industry.

The question of how to prevent a flood of imported oil products if the prices of foreign and domestic crudes are equalized remains with us. The basic answer is not to permit the equalization of foreign and domestic crude oil. Decontrol is not in the interest of domestic non-integrated refiners. Most domestic refiners, however, are also oil producers, so profits from decontrol would overwhelm any possible reduced income from refinery operations.

The halting of phased deregulation and maintaining of the entitlements program would retain the present entitlements benefits for imported crude and prolong the anomaly of subsidizing imports using entitlement funds paid in by the domestic producers.

The answer is the phased elimination of import entitlements benefits, converting the entitlements program to a purely domestic operation. The present higher profitability of foreign crude because of its more favorable tax treatment and the other factors discussed earlier would be ended. Taking the profits out of the importing of oil is the market approach to reducing our heavy dependence on imported oil.

Under the above policy, however, there would be no incentive to import foreign crude and to refine it here instead of importing foreign refined products. The solution would be to establish a fee on oil products sufficiently high to discourage imports. This would be probably at about the level proposed by Senator Johnston. Admittedly, the fee would raise the price of all oil products, foreign and domestic alike. The savings to consumers through the retention of price controls on domestic crude would far more than offset the costs of the import fee on oil products.

Presentation by Herschel F. Clesner

As I look at and listen to the group of panelists, everyone represents a vested interest. If I could represent someone who is left out, that would be the ultimate user or consumer. Thus I will, first, make comments and put some thoughts into perspective and, second, make a brief presentation on behalf of the public or the consumer. As a matter of fact, I will attempt to serve as a "people's counsel" before a regulatory agency. In doing so, I would like to first add this disclaimer, that whatever views I express are my own and not those of the House Subcommittee on Commerce, Consumer and Monetary Affairs, of which I am chief counsel.

One of the first things that comes to mind is simply this: It appears that some speakers are attempting to find out just what our refinery policy is, while others are attempting to evaluate what they deem it to be. At best, we have heard that if a policy exists, it is unreliable. Yet energy is the most important issue before this country today and, of course, refinery policy is tied to and an integral part of it.

Do we really have an energy policy, or do we have contradictions? Do our energy policymakers have any input into energy tax policy? The hearings of the Subcommittee have shown conclusively that they do not. And yet 60–70% of our so-called energy policy is in the tax field.

Do we do justice to our economy, to our people, and to the consumer, and is there equity in this process?

The word "equity" can be used in a number of ways, like a lot of other words in this area which are slogans, canards, and catch words. And in terms of "equity" the panelists keep referring to "capital intensive industry" and "investment." Yet today it is a changing picture out there industrially, and particularly with respect to costs.

Now let us consider this: We find activities are not capital-intensive today to the degree that they were because capital, labor, interest, and operating expenses amount to about 6–7¢ per gallon today. Crude oil acquisition costs are by far the most dominant factor, at 57¢ a gallon. Concession agreements are, therefore, the most important variable in the refinery process to a domestic refiner as well as a foreign refiner; capital costs, inasmuch as supply agreements can attract anything, are another factor; and recessionary capital, at 8% of cost, is a minimum consideration.

As to small refiners and their needs, they are in a bind. But the mechanisms or the approaches that have been suggested are Rube Goldbergesque. With no definite guidelines and no policy, favoritism takes hold.

In considering the multinationals, I have heard the word "we" dealing with such a refinery policy where it concerns that which operates abroad as well as that which operates domestically. Is that a policy dominated by and run for the benefit of the multinationals, our five members of the so-called Seven Sisters, rather than the U.S. public and other refiners?

I think what the other panelists have to say is meaningful as to the Administration, but to equate their needs and their requirements with those of the public is misleading because they are not always the same.

And in this changing world it is not merely OPEC going downstream by building refineries, making consignment agreements. It is about time that we looked at the Caribbean, as well. It is possible that a number of those countries with refineries may invite us to go home. Yes, it is time that we looked hard at the picture of security, even in the Caribbean. The Florida straits can be dominated just as well as the Straits of Hormuz. Though we are not the security panel of the conference, there is a complete overlap with economics.

Furthermore, as to the availability or lack of availability of crude oil, the discussion here about the question of light, sweet crudes has centered on the OPEC nations. There has not been any consideration of the growth in the last few years of light crudes on the world markets. Mexican production has doubled, and may well quadruple in another two years depending on government policy, to an amount of possibly three million barrels a day more than ever existed.

Nigerian, Angolan (Cabinda) and North Sea crude production is way up and is sweet and light, too. There is an overall increase. Alaskan Slope production is also way up. When we talk about heavy and sour, there has been a glut for 10 years.

In 1974, immediately after the embargo, fiscal incentives were given so that we would have investment within the United States to modify and retrofit refineries to handle more heavy and sour crude. Companies declared investment programs. But if we go back and take stock, very little, or let's say only a small percentage, was put into effect by the companies.

What has happened in this country recently?

I would like to summarize the findings of an analysis made by the Subcommittee, which I am submitting for inclusion after my remarks in the conference record.

In a 13-month period, refinery product prices have soared skyward to outer space. The profits of the oil companies each day have reached new unbelievable heights and continue to reach into the cosmos. At the refinery level, the increase in the average price of refined products is going up far more rapidly than possibly can be justified by increases in crude oil costs to the refiners.

What our analysis showed was essentially this: The national average of wholesale price increases for heating fuel and diesel fuel between September 1978 and September 1979, the lastest period for which full data are available, indicates that domestic oil refiners' profits increased by more than 800% for heating fuel and 700% for diesel fuel during this 13-month period. The American consumers have been overcharged as a result of these profit increases by more than $3 billion for heating and diesel fuel purchases. And the distillate fuel oil price rises were way out of proportion to increases in the price of crude oil.

These unconscionable price and profit margin increases can be blamed, in significant part, on lax governmental enforcement of the price standards for distillate fuel oils by the Council on Wage and Price Stability. The Council's failure to enforce these price guidelines against our domestic refiners relates to the fact that they only use four COWPS employees, part time, not full time, to monitor and enforce compliance with the price standards in a $400 billion a year industry.

COWPS' responsibility for energy prices must be shared with the Department of Energy, an agency whose national energy goals are based on higher, not lower, energy prices. The Council's guidelines are difficult to enforce because they apply to the total petroleum product of the firm, not the specific products such as heating oil and diesel fuel. And until very recently, COWPS' price guidelines encompassed a firm's non-energy as well as its energy-related activities.

America's fight against inflation depends to a significant extent on the ability of the Council on Wage and Price Stability and the Department of Energy to enforce their price standards for petroleum products. Thus far, these agencies have apparently failed in that important mission.

Now, how did this come about? We have heard that there is a shortage out there because of an Iranian shortfall. Well, the Subcommittee checked the U.S. Bureau of Census' published U.S. Customs figures for crude oil imports for the first five months of 1979. The increases were 10% over 1978. Someone said, "Compare it to 1977. That is the high year." You bet your life it was the high year. And you know what? In 1979, after five months, the difference was three-tenths of one percent.

What has happened is this: The average citizens cannot eliminate the necessity of energy from their lives, and their resultant plight has moved to a state that can be best described as pitiful. It is time that their state of mind be taken into account because inflation is running in excess of 14% annually and energy costs at an annual rate of 43%.

What do we need? Yes, we do need an energy policy. We need a refinery policy, one that is clear and definitive, one that takes into account our national security and the economics of this country, one that does not grant foreign crude oil credits or benefits over U.S. crude, one that will not benefit foreign refining operations at the expense of domestic operations, one that does not indulge in recycling through Rube Goldberg-type mechanisms.

Moreover, in furtherance of an energy and refinery policy, we should definitely do what we can to expand the development of heavy oil, which we have in great abundance. Heavy oil has been neglected. It deserves to be looked at hard and promoted. Obviously, fiscal incentives should be encouraged.

Analysis of Recent Price Increases for Distillate Fuel Oils Indicates Failure of the Council on Wage and Price Stability to Enforce Price Standards for Petroleum Products

Prepared by the Technical Support Staff, House Subcommittee on Commerce, Consumer and Monetary Affairs

In connection with the Subcommittee's continuing investigation of the Council on Wage and Price Stability's enforcement of the price standards for petroleum products, please find below an analysis of heating fuel oil and diesel fuel wholesale prices charged by domestic refiners:

Objective

The wholesale price of distillate fuel oils (specifically of diesels No. 1 and 2, and heating fuel oils No. 1 and 2) has practically doubled over the last year. At hearings held by the Subcommittee last year Dr. Barry Bosworth, then Director of the Council on Wage and Price Stability stated, ". . . at the refinery level the increase in the average price of refinery products including gasoline, home heating oil and others, is going up far more rapidly than can possibly be justified by the increases in crude oil cost that are going to the refiners."[1] However, in the field of fuels and related products the Administration's efforts to control prices are at cross purposes with its more general policy of promoting conservation (and hopefully encouraging additional domestic production), by increasing consumer prices to levels commensurate with world prices.[2]

The objective of this study is to analyze the change in the national average wholesale prices of heating fuel oil and diesel fuel for the past year. Another objective is to determine whether the increased prices fall within the wage and price standards. Lastly the study will identify price components which have contributed to these pricing changes.

Methodology

To determine the causes of the domestic price increase, it was necessary first to identify, and then to quantify in a monetary sense, the input factors to heating fuel oil and diesel fuel costs. Data on these factors had to be obtained for the time period in question, and each factor analyzed alone and in combination in order to determine those forces which contributed most significantly to the price rise.

Source of Data

The data used in this study were published in the Department of Energy's (DOE) "Monthly Energy Review" and "Sales of Fuel Oil and Kerosene in 1978" and in the National Petroleum Council's (NPC) "Refinery Flexibility" report. Numerous contacts were made within DOE to verify published figures, and to obtain updated figures and other statistical input. In order to obtain consistent and complete data for each of the variables involved and in an effort to keep the analysis as timely as possible, the data series employed run from September 1978 through September 1979, comprising a 13-month period.

Analysis

• Wholesale heating fuel oil prices increased from 36.9¢ per gallon in September 1978 to 69.0¢ per gallon in September 1979.[3] The difference of 32.1¢ per gallon represents an increase in price of 87% during this period (see Table 1).

• Wholesale diesel fuel prices increased from 37.1¢ per gallon in September 1978 to 69.0¢ per gallon in September 1979.[4] This difference of 31.9¢ per gallon represents an increase in price of 86% during this time period.

• The biggest single cost factor for refiners is their crude oil acquisition cost, which amounted to 29.9 ¢ per gallon in September of 1978 and increased by 18.1¢ per gallon, to 48.0 ¢ per gallon in September of 1979.[5] This represents an increase in cost of 61% for this period (see Table 1).

• The difference between the wholesale heating oil price from refiners and the refiners' average crude oil acquisition cost is the refiners' margin. This margin was 7.0 ¢ per gallon in September 1978. By September 1979 it had climbed to 21.0 ¢ per gallon. The 14.0 ¢ per gallon price jump represents a margin increase of 200% for heating fuel oil (see Table 1). This is out of line with historical trends (see Table 2).

• The difference between the wholesale diesel fuel price and the refiners' average crude oil acquisition cost is also the refiners' margin. This margin was 7.2 ¢ per gallon in September 1978. By September 1979 it had climbed to 21.0 ¢ per gallon. The 13.8 ¢ per gallon price increase represents a margin increase of 192% for diesel fuel (see Table 1).

The refiners' margins can be broken down into two basic components: costs and profits. The NPC study reports operating costs of 5.4 ¢ per gallon in 1978[6] (see Tables 3 and 4). This figure includes costs for labor, capital, interest, operating expenses, etc. Profits are a residual of the refiners' margin less operating costs which result in a net profit of 1.6 ¢ per gallon for heating fuel oil and 1.8 ¢ per gallon for diesel fuel in September 1978.

The refiners' margin for heating fuel oil had increased from 7.0 ¢ to 21.0 ¢ per gallon by September 1979, and for diesel fuel had increased from 7.2 ¢ to 21.0 ¢ per gallon for the same period. Average operating and capital costs have remained relatively constant. Prevailing inflation rates were 11–12% during the period; the industry had undertaken only modest capital expansion; and labor costs had increased at less than the rate of inflation under the Administration's wage/price control program. Because of relatively higher costs for fuels used at refineries,[7] we assume an overall cost increase of 20%[8] from the NPC reported costs of 5.4 ¢ per gallon. This calculation yields an estimated cost average for 1979 of 6.5 ¢ per gallon and, consequently, a profit figure of 14.5¢ per gallon for heating fuel oil and diesel fuel. The net refinery profit jump from 1.6 ¢ and 1.8 ¢ to 14.5¢ per gallon (allowing for operating cost increases) works out to a relative profit increase in excess of 800% for heating fuel oil (see Table 3), and over 700% for diesel fuel (see Table 4).

Accordingly, it is estimated that the increased profit margins for these middle distillate fuels yielded refiners an additional $3.4 billion profit.[9] This additional $3.4 billion cost to the American consumers is the result of refiners' net margin increases of 12.9 ¢ and 12.7 ¢ per gallon for heating fuel oil and diesel fuel respectively, over and above the profitability the industry enjoyed before this period. The volume involved consists of estimated U.S. deliveries from domestic sources only of 53.2 billion gallons[10] for the period September 1978 through September 1979. This sudden increase in domestic refiners' profit margins has contributed substantially to inflation and violates the spirit of the wage/price guidelines. None of the COWPS pricing standards has been able to effectively restrain the dramatic price increases and the resultant profit margin increases for these middle distillate fuels.

Conclusions

- Domestic refinery profit increases from September 1978 through September 1979 amounted to over 700% for diesel fuel and over 800% for heating fuel oil.
- American consumers of these middle distillates have involuntarily contributed more than $3 billion in unwarranted refiners' profits because of the absence of price standards for middle distillates.
- None of the COWPS pricing standards has been able to effectively restrain the dramatic profit margin increases for these middle distillate fuels.

TABLE 1
Crude Oil Cost and Refinery Wholesale Price for Distillate Fuel Oil
1978–1979
(in cents per gallon)

Month	Average Crude Oil Acquisition Cost	Heating Oil		Diesel Fuel	
		Average Wholesale Price	Refiners' Margin	Average Wholesale Price	Refiners' Margin
1978					
September	29.9	36.9	7.0	37.1	7.2
October	30.0	38.1	8.1	37.7	7.7
November	30.4	39.4	9.0	38.6	8.2
December	30.8	40.1	9.3	39.1	8.3
1979					
January	31.2	40.9	9.7	39.7	9.5
February	32.0	43.1	11.1	41.8	9.8
March	32.6	45.8	13.2	44.5	11.9
April	34.6	48.3	13.7	47.7	13.1
May	36.7	53.2	16.5	53.4	16.7
June	40.5	58.8	18.3	58.7	18.2
July	44.2	62.5	18.3	62.4	18.2
August	47.0	65.7	18.7	66.0	19.0
September	48.0	69.0	21.0	69.0	21.0
INCREASE Sept. '78 - Sept. '79 cents per gallon	18.1	32.1	14.0	31.9	13.8
%	60.5	87.0	200.0	86.0	191.7

Source: Department of Energy, "Monthly Energy Review," November 1979, pp. 80, 86, 88. Direct communication with Department of Energy officials.

TABLE 2
Heating Fuel Oil Refiners' Margin
1976–1979[1]
(in cents per gallon)

Year	Month	(Average Heating Oil Wholesale Price) −	(Average Crude Oil Acquisition Cost) =	(Refiners' Margin)
1976	Sept.	31.1	26.4	4.7
	Dec.	33.7	27.0	6.7
1977	Sept.	35.5	28.6	6.9
	Dec.	36.6	29.2	7.4
1978	Sept.	36.9	29.9	7.0
	Dec.	40.1	30.8	9.3
1979	Sept.	69.0	48.0	21.0

[1]Until June 1976 distillate fuel oils were under Federal price controls.

Source: Department of Energy, "Monthly Energy Review," November 1979, pp. 80, 88. Direct communication with Department of Energy officials.

TABLE 3
Net Refiners' Margin for Heating Fuel Oil
September 1978 - September 1979
(in cents per gallon)

Month & Year	(Average Refiners' Gross Margin) −	(Average Refiners' Operating Costs) =	(Average Refiners' Net Margin)
9/78	7.0	5.4	1.6
9/79	21.0	6.5	14.5
% change 9/78 - 9/79	200	20	806

Source: Department of Energy, "Monthly Energy Review," November 1979, p. 88, derived. Direct communication with Department of Energy officials. National Petroleum Council, "Refinery Flexibility," November 1979, p. 90.

TABLE 4
Net Refiners' Margin for Diesel
September 1978 - September 1979
(in cents per gallon)

Month & Year	(Average Refiners' Gross Margin)	− (Average Refiners' Operating Costs)	= (Average Refiners' Net Margin)
9/78	7.2	5.4	1.8
9/79	21.0	6.5	14.5
% change 9/78 - 9/79	192	20	. 706

Source: Department of Energy, "Monthly Energy Review," November 1979, p. 86, derived. Direct communication with Department of Energy officials. National Petroleum Council, "Refinery Flexibility," November 1979, p. 90.

TABLE 5
Sales of Distillate Fuel Oil by Use as Percent of Total
(millions of barrels)

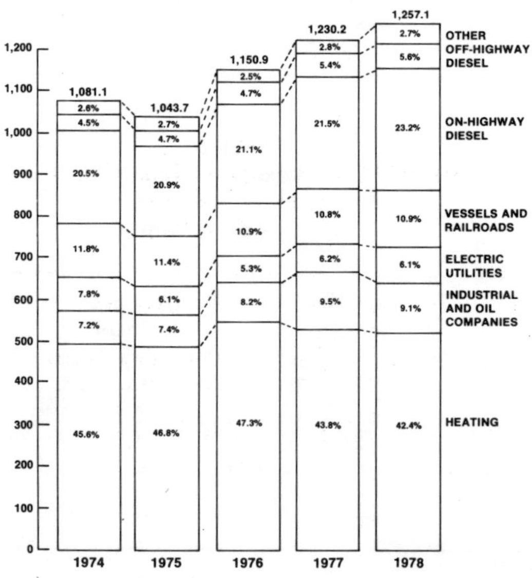

Source: Department of Energy, "Energy Data Reports—Sales of Fuel Oil and Kerosene in 1978," p. 2.

References

[1]Subcommittee on Commerce, Consumer, and Monetary Affairs of the House Government Operations Committee, Hearings on "Adequacy of the Administration's Anti-Inflation Program, Part 2," May 3, 4, 5, 7, 11, 18; June 18 and 28, 1979, p. 850.

[2]*The National Energy Plan*, Executive Office of the President, April 29, 1977, p. XI–XII.

[3]Department of Energy, "Monthly Energy Review," November 1979, p. 88 and direct communication with DOE officials.

[4]Ibid., p. 86.

[5]Ibid., p. 80.

[6]National Petroleum Council, "Refinery Flexibility," Vol. I, (an interim report) November 1979, p. 90. The NPC interim report is the result of surveys conducted in early 1979 of 174 companies operating 289 refineries in U.S. jurisdictions. Total response represented 16.9 million b/d of U.S. refinery capacity or 98% of the total. The average operating cost figure is $2.29 a barrel. Almost half of this cost is for fuel and purchased utilities, with depreciation, maintenance, payroll and administration accounting for the remainder. The certified public accounting firm of Arthur Young and Company collected and aggregated the data, preserving the confidentiality of individual company data.

[7]The increased costs to refiners would be less than the wholesale price suggests; however, due to the fact that refiners' profits cannot be counted as a part of refiners' increased costs.

[8]This is apart from the refiners' crude oil acquisition cost for throughput. The 20% increase is an estimate based on the Nelson refinery operating cost index of 18.5% increase between August 1978 and August 1979.

[9]DOE/EIA, "Sales of Fuel Oil and Kerosene in 1978." Of the total distillate fuel oil category delivered in 1978, Table 6, p. 6 identifies heating fuel as comprising 42% of the total. Table 2, p. 4 also identifies diesel fuel used on-highway and off-highway as comprising 29% of the total distillate fuel oils. The remaining 29% of distillate fuels consists of either heating fuel or diesel used primarily in industries for which the government does not collect specific information.

[10]Based on 1978 experience. See DOE, Petroleum Statement, 1978 (Final Summary), Table 1, p. 2 and adjustments.

DISCUSSION

MR. MAGEE: I feel somewhat obligated to comment on the 800% increase in profits on heating oil, and I would also like to state for the record that I am speaking for myself, not my company, unless otherwise specified.

My company does share the concern expressed regarding the rising level of prices to the American consumer. Having been involved in numbers all of my working career, I think it is important to recognize that given the phenomenal abundance of numbers today, you can put together almost any set to make any story you choose.

It is absolutely critical that, when looking at the margin in the refining business, we look not at the difference between the price of an individual product, such as heating oil, and the cost of its major component, crude oil, but rather at all the products made from a barrel of crude oil, since when you run the crude oil you do not make barrel for barrel heating oil. You are required to make other products, and the marketplace is likely to assign different values to those other products.

Currently, over 30% of the product made in the United States is neither gasoline nor distillate but, rather, products such as residual fuel oil, which are being sold at prices less than the cost of the crude oil required to make them. If you will look at that you will see a somewhat different story.

I cannot comment on industry statistics. I do know that in comparing the one month of September 1979 to the one month of September 1978, the gross margin as used in the study for Atlantic Richfield Company, when you look at a barrel of finished product, was actually lower in 1979 than it was in 1978. If you look at the third quarter of 1979 versus the third quarter of 1978, there was no change in our gross margin.

It is very important not to look at heating oil in suspended isolation because to date nobody has come up with the equipment to make just heating oil from a barrel of crude.

MR. O'NEILL: Let me make two comments.

As one who spent the last 31 years scurrying around trying not to get stepped on and squashed by the international majors, my most dominant feeling toward them is probably fear, but once in a while it is sympathy.

And as to these profits that have been described as obscene, astronomical, and so on, I think we ought to be fair. Their profits, measured as a percentage of gross revenues, as a percentage of their investment, are less than those of *The Washington Post,* ABC, NBC, CBS, and many others. I think that they are taking a bad rap there.

The other comment I would make with regard to consumers: I think the rest of the world views the American consumer as a profligate consumer of energy. And I think our grandchildren will castigate us for the manner in which we have consumed energy.

I am very much in favor of all sorts of conservation in this country. I would favor, for example, a 100% sales tax on gasoline. Well, that is not going to happen because the consumers are not going to permit it—you and I, and I am just as bad about it as anyone else. But I don't feel too sorry for us when I see what goes on around the rest of the world.

MR. MALIN: Thank you, Mr. O'Neill. In these troubled times we need all the help we can get.

I would like to make two remarks, one following up on Mr. Magee's remark about the Rosenthal study. There are some things missing from that study. It suggests that refinery margin after reduction by refinery operating cost is profit. But the refinery margin itself is determined by taking the difference between crude costs and the price of distillate oil sold from terminals or bulk plants downstream of the refinery.

If you think about it you will note that such costs as marketing costs, distribution costs, bridging costs—which means getting the product to be sold from the refineries to the downstream plant or bulk plant—overheads, and taxes are among those costs that are missing from the definition of refinery margin. These are real costs; these are cash costs. In fact, given the level of "refinery margin" identified in the study, which I think you will find to be 1.5¢ a gallon in 1978, if you take some of these other costs off, you will find that heating oil sales on the average for the industry in September of 1978 may well have been sold at a loss.

Therefore, to simply use "refining margin" as defined in the study is an inadequate measure.

I would support and leave to your imagination the remarks that Mr. Magee has already made with respect to dividing this cow we have called crude oil into its various segments and determining that the cost of making steak and the cost of making hamburger are the same.

The second comment I would like to make is that notwithstanding the fundamental disagreement that we have on this panel perhaps, and certainly in the nation at large, on the disposition of the revenues from higher prices—who gets what and how much and what they do with it—I think that we ought to try to understand that it is important that as a nation we permit our domestic prices to rise to meet those of the alternates, which are world crude oils. To do any less is to subsidize demand in this country, and that demand must be met by incremental imports of foreign oil.

It seems to me this confounds every objective of energy policy.

MR. CLESNER: If you look at the study you will find we did build in overall costs, drawn from the National Petroleum Council's reported costs, 5.4¢ a gallon, and then worked up to a further inflationary increase as described in both Footnotes 7 and 8; then we took the higher cost figures by the industry's own Nelson Refinery Operating Cost Index, which came to an 18.5% increase between August of 1978 and August of 1979.

However, for that period, instead of settling for 18.5%, we took 20%. So I would say we were conservative because the base figures are still there. The factors are basic constants in September 1978 and 1979. The difference is expressed in the additional 20%.

Furthermore, as expressed by the Chairman of the Subcommittee, the invitation is open to industry to refute the analysis. If they dispute these figures, the Subcommittee would be glad to have them come forth with their own figures and testify at hearings. The door is open: We would not like the record to remain clouded for the purpose of report-writing or anything that is done by the Subcommittee members in considering the problem.

MR. COLLINS: There is a note of political unreality which underlies a great deal of discussion here. I will submit that one will not learn anything from the industry statements beyond what you read in the Mobil ads, the Chevron U.S.A. ads, the Texaco ads. Those have almost become dulled in reading, they are coming out with such frequency saying essentially the same philosophical concepts. I have come to the conclusion in talking to a large number of people in the street, including our own membership, that these advertisements have become counterproductive.

The fact of the matter is that we have political problems that the oil industry imagines can be solved by their muscle on Capitol Hill. I am very disturbed about the question of a proper refinery policy. I don't know for sure what a proper refinery policy should be. What is going to happen is that next year, after the congressmen and senators have been home running for office and talking to their constituents, they will find that there is an unparalleled dislike for the oil companies at the present time. When we come asking for $1.26 per barrel import fee on oil products to save the domestic industry, of which we are a part, we are likely to get a biased response.

What we would suggest is a trade-off. The threat of a flood of imported refined products in a couple of years is real. What we need to do is preserve a price control on domestic crude arranged at an equitable level. To increase prices of old oil all of a sudden from $7.00 per barrel, remembering that the $7.00 isn't all profit, that the real profits can be counted at about $3.00, you are then changing profits from $3.00 to around $23. This is something that oil company people forget, that people can add, subtract, and take percentages and talk about this.This is something that white collar people do not understand about people who work with their hands. They are just as smart, I have discovered, as people who have been to college. And these people all have votes and they will be talking to their congressmen.

I am afraid unless the domestic refining industry works out a trade-off with the public so that the public gets something out of the deal, nothing much is going to happen.

MR. McGREGOR: I would first like to respond to what Mr. Collins just said. Although I understand this is a refinery policy forum, I should point out that he is absolutely right when he says we are on a phased decontrol schedule for domestic crude production. He has left out the fact that flowing crude oil will most likely be taxed under the windfall profits tax at 70% per barrel. It will probably boil down to something between $12.75 and $13. So the public is definitely picking up a good portion of that net.

Turning to the issue at hand, I think everyone participating in the seminar, from the major independent oil companies to Mr. Clesner as the "people's counsel," will all agree that refinery policy in the past has been formulated on an ad hoc, sporadic, and often counter-productive basis. Hindsight is good. Perhaps we would have been better off just to regulate refineries beginning in 1973 on a public utility type of regulatory concept. We would have had the bureaucratic expertise to do that. That is no longer the case.

We are looking forward now to the decontrolled environment. It is important to keep that in mind; that is our baseline. Any changes which are adopted either administratively or legislatively should be done with substantive justification and minimization, if not elimination, of special and/or political-type interests which are counter-productive to our nation's energy policy goals.

RAY F. BRAGG (American Petroleum Refiners Association): I'd like to pose a question to Mr. O'Neill. You touched on crude access at fair prices. I wonder if you would expand on the impact which the failure to follow through on that would have, not only upon you as a refiner, but upon the markets that you serve.

And, secondly, since you are an individual who has been involved with the development and building of a refinery of 200,000 barrels and are presently operating a 30,000 barrel refinery, perhaps you might have some thoughts of the differences in economies of scale between large and small refinery operations.

MR. O'NEILL: Well, I believe that much too much has been made of the economy of scale. I think most people have the idea that the larger a refinery is, the more efficient it is. I don't agree with that. My own opinion is that probably the optimum size is between 50,000 and 150,000 barrels a day; I mean, 200,000 barrels a day might be a bit off.

But size is only one factor that has to do with the efficiency of a refinery. There are many others that need to be considered.

One of the important ones is when it was built. To be sure, a lot of old refineries have been revamped many times and changed and managed to keep up with the times to a very reasonable degree, but they are not as efficient as one that is brand new. And this is particularly true in thermal efficiency.

When we first started to design the ECOL plant, it was before crude prices shot up as a result of the embargo, and we were using 75 ¢ a million BTU for thermal costs. We changed that to $2.00 per million BTU, making the plant more efficient from a thermal standpoint. A good example is Texaco. They have a very, very large refinery at Port Arthur, but they have a number of smaller refineries around the country. I am satisfied that they have studied each of those in great detail and arrived at a decision in those cases. Size is not the only factor which is important.

A refinery without crude is no good at all. And independents who have no production of their own are at a great disadvantage. Most people have built where they can get access to foreign crude. Unfortunately, the original aim of the entitlements program to equalize the cost of foreign crude which is on a free world market, and the controlled price of domestic crude, has not worked. A few years ago the spread was 40–60 ¢ a barrel. Now it is much greater than that, as you know.

The other problem, though, has to do with the subsidies for production, if you want to call a tax incentive a subsidy. That is a very difficult one for me to deal with because I think the people who are in exploration and production very badly need those incentives. But it does give them an area where they can have an advantage over a company that has no production of its own, both as to access to the crude and cost of the crude. I am not quite sure what the answer to that is.

MR. FRED SCHULMAN (Consultant): My question is to Mr. McGregor and Mr. Malin.

First of all, it relates to decontrol and the question of a proper price for alternate fuel.

It seems to me the word "decontrol" in this sense is really confusing and misleading. We are really not talking about decontrol but about transfer of control from here to OPEC. In other words, the decontrolled price a year ago would have been roughly $12.50, then $14.50, and $19, and now it is $26 and on its way to $30. So a decontrolled price is a price over which we have had no control. It is still controlled but it is not controlled here. For example, if Libya decides to raise the price by $4.00 a barrel, that is the new controlled price.

This relates to the question of alternate fuels in the following way, and I wonder what is your reaction to it:

If you recall, way back at the time of the embargo, shale oil as an alternate fuel was estimated to cost between $5.00 and $7.00 a barrel. And it keeps going up and up and up along the lines you all indicated.

Now as long as the Department of Energy and the rest of us are unwilling to put some restraint on OPEC's ability to raise prices, which we can do by the mechanisms you have talked about and others, then all that has to be done to defeat alternate fuels is for OPEC to keep raising prices—the cost of the alternate fuels becomes astronomical and plants never get built.

MR. McGREGOR: I don't think it can be said that up to now price controls have served to subsidize OPEC's prices. Therefore, the marginal barrel has come from OPEC, and OPEC has the pricing authority over its own production. There will be a demand response from decontrol, and to the extent that means less demand for the imported foreign barrel, I think it puts our energy future more into our hands and less into the hands of OPEC.

Second, with respect to industry estimates on alternative fuels, the costs of producing, the costs of research and development, and so on, of alternate fuels, have certainly increased over time. I think you have to throw in an inflation component. But I agree there is some cut-off point regarding the OPEC price. Right now at DOE we are developing what we call a premium component, which is X dollars per barrel we'd be willing to pay on top of the prevailing OPEC price in order to foster alternative domestic production. That premium is probably no greater than $5.00 a barrel. I think you will find that as policies are evolved out of DOE for the future, they will be based on that type of premium concept.

MR. MALIN: It is important to remember that OPEC is a reality. They have the crude oil, and until we can in effect say, "we don't need the crude oil," through conservation or demand constraint or alternate sources of energy—whether domestic crude oil or some sort of alternate energy—we've got to deal with OPEC where it is.

I suggested earlier that it is important, at least in my perception, that we insure we are getting the maximum domestic benefit in terms of conservation and incentives for alternate energy resource development by charging ourselves the going price for our current alternate, which is OPEC oil.

Now, I think we are beginning to see, absent any other program, encouraging developments for alternate energy in this country in terms of investment. It is going to take some time to develop alternate energy, there's no question about it, and in the intervening time I am not sure I see where our energy is coming from, unless it is from OPEC.

MR. STEPHEN GOLDBERG (Office of Senator Bentsen): If we let things go the way they are planned to go, decontrol will come on the books in October 1981, and the Emergency Allocation Act also expires about the same time. Considering the fact that we have had a series of grassroots refineries pop up over the past four or five years, heavily dependent upon sweet oil to process, and further assuming these same refineries will not retrofit between now and October 1981 and will therefore continue to be heavily dependent on this sweet oil—which is in shorter supply than other crude oil—what does the panel propose be done in allocating the sweet oil between majors and the independent refiners?

MR. McGREGOR: In my presentation I outlined the facets of the crude oil access problems we are looking at. I can only go back to my statement and say we are looking at them, and we realize it is a serious problem, and hopefully something will be done well in advance of October 1981.

I should point out, however, that the standards of measurements are not only for equity considerations for keeping existing refineries in existence, but also for economic efficiency, cost to consumers, and certainly the national security component, as well as the employment and balance-of-trade issues.

MR. MAGEE: Well, it is a question that I don't have an answer to, other than to point out that Mr. O'Neill indicated the efficient refinery is one that has the capacity to process between 50,000 and 150,000 barrels per day of crude oil. Thirty of the 35 refineries that came on stream recently have a capacity of less than 50,000 barrels a day; they are largely refineries that have the ability to process sweet crude. The government, as Mr. McGregor indicated, is considering the policy issue of whether we want to continue with such things as the small refiner bias that are going to drive people to this group.

MR. O'NEILL: We have never supported the small refiner bias. We built our refinery thinking the bias would be gone by the time our plant came on stream. It wasn't. The only thing I can say in support of continuing the small refiner bias is that I don't think there is any question but that it induced a large number of people to make a financial investment and, in fairness, I think the government has some responsibility to some extent to those people. I think it was a mistake to have the small refiner bias in the first place, and I don't know anyone who disagrees with that.

MR. MALIN: Let me refer back to the comment I made at the very end of my prepared remarks, that is that it is an enormously complex problem. It probably goes beyond October 1, 1981. It is probably the fundamental question to be addressed in the 1980s.

But as I suggested in my remarks, we have to be sure whether we are talking about a supply problem or a price problem. And within the supply problem, is the problem one of inadequate supply or is it one of access? And with respect to the pricing problem, is it one of disparity in pricing between sources or is it one of being just plain too high—the "how much is too much?" sort of question.

But the theme of my remarks, if there was one, was that we have to get far enough out in the 1980s to be able to look back and say with respect to the demand that we want to meet from our refineries, "Do we have the capability? Is crude adequate?," and so on. We can then address what sort of mechanical emergency program we might wish to have to deal with an emergency problem. But the long-term issue is a very serious question. I don't pretend to have the answer, but it is a fundamental question.

MR. CLESNER: I believe it is a very fundamental question, one that will have to be looked into in the '80s, based on the rationale pointed out. But further than that, it clearly comes down to this: Once, having urged these people to invest, does the government intend to insure equity for them? Will demand then be cut, for example, from the majors? Will the sweet crude oil that the small refiners are allowed by the government to buy be at a high price, an average price, or a low price on the market? That is an important question because the price is different today.

The difficulty has become compounded because one producing nation after another is going the way of attempting to make country-to-country agreements. Whether that is good or bad makes no difference in this context. It just happens to be a fact which must be acknowledged and which means the present method of obtaining or distributing foreign sweet crude may not be pertinent in the 1980s.

But having looked at that, somewhere a policy decision has to be made, and it must go to the equity of the investment, the difficulties of the times, and the prices.

MR. JOHN LAMONT (Lobel, Novins and Lamont): First, I'd like to comment on something said by Mr. O'Neill and Mr. Malin. They talked about the profit by the American consumer in the use of crude, and Mr. Malin noted particularly that until we permit the domestic price to rise to the world price, the highly artificial world price, we are in effect subsidizing consumption. I wonder if he recalls that for about 40 years we maintained a price domestically about 30% higher than the world price in order to foster development of domestic production, a process in which Mr. Malin's company profited quite greatly.

MR. MALIN: I am not going to defend the earlier protectionist policy except to suggest there were a different set of circumstances that we were trying to deal with at that particular point.

MR. LAMONT: But it did tend to discourage consumption.

MR. MALIN: By higher prices, yes.

MR. LAMONT: Yes. So we have in essence followed that policy of discouraging consumption since the early 1930s.

MR. MALIN: Well, perhaps since the early 1930s, but we certainly reversed it in 1974.

MR. O'NEILL: Until the late 1950s the world price of crude oil and the domestic price of crude oil were substantially the same, around $3.00 or $3.50 a barrel. When the independents broke the monopoly of the Seven Sisters by discoveries in Venezuela and then Libya and then in the Persian Gulf, there came a great flood of foreign oil into the United States, and then domestic prices were propped up well above the international price.

Now, that was done to encourage continuation of exploration and production in the United States. I don't think it had anything to do with encouraging people to conserve.

To go back to the early 1930s, when the Texas Railroad Commission was established, that was a conservation effort.

MR. LAMONT: May I just correct it as a matter of fact. The flood you spoke of started in 1954, not in 1960.

MR. O'NEILL: I said the late 1950s.

MR. LAMONT: You had tax credits beginning about that time which gave you a good reason for it. But it was the rationing system, plus the import controls, which maintained the price of American crude at least 25–30% above the going price of world crude at any point from about 1934 until about 1970.

Then, as Mr. Malin observed, it changed.

MR. O'NEILL: Well, as an independent refiner, we were unable to buy foreign crude below domestic prices until about 1955.

MR. LAMONT: My second question concerns Mr. McGregor's statement that DOE does have under consideration various means of equalizing access to the foreign crude as far as the domestic independent and other refineries are concerned.

Now, I presume that in the consideration for that, the mechanism by which that equalization will take place will be some distortion of the taxing system or of the remaining regulatory system, or DOE will set up a new foreign entitlements program entirely out of whole cloth. But when they have done that, haven't they in essence said that they cannot decontrol crude oil because the entitlements program has truly been the regulator of the price of domestic crude oil?

MR. McGREGOR: Mr. Lamont failed to recognize one of the options of the statement, and that is to allow the market to set the framework as to who has access to crude oil. I pointed out we were aware of the major refinery policy issue and are in the process of analyzing it right now. You are aware a subsidy has been allowed in certain situations where crude access has become a problem to certain companies in the very recent past. That is also our policy interest.

MR. KIRBY BRANT (House Subcommittee on Commerce, Consumer and Monetary Affairs): I want to ask Steve McGregor if I was right in understanding him to say that consideration is being given to the amount of premium we ought to be willing to pay *over* the world price in order to be assured of a nationally secure supply.

MR. McGREGOR: Yes, we are doing that analysis.

MR. BRANT: I think that is perfectly analogous to the old mandatory oil program. That, too, was justified on the ground of national security.

MR. McGREGOR: There is a national security component in the premium analysis, and in that sense they are analogous, but I think the special interests versus the public interest would be totally different.

Abbreviations in this Volume

ARAMCO: Arabian American Oil Company
ARCO: Atlantic Richfield Company
b/d: Barrels per day
COWPS: Council on Wage and Price Stability
DOE: Department of Energy
ECOL: Energy Company of Louisiana
EIA: Energy Information Administration
EPA: Environmental Protection Agency
EPAA: Emergency Petroleum Allocation Act
FEA: Federal Energy Administration
GAO: General Accounting Office
IEA: International Energy Agency
LNG: Liquefied natural gas
LPG: Liquefied petroleum gas
NGL: Natural gas liquids
NPC: National Petroleum Council
OCAW: Oil, Chemical and Atomic Workers International Union
OPEC: Organization of Petroleum Exporting Countries
OSHA: Occupational Safety and Health Act
PAD I, II, III, IV & V: Petroleum Administration for Defense
  Districts. The nation is divided into five PAD districts.
SEC: Securities and Exchange Commission
SPR: Strategic Petroleum Reserve

About the Editors

Bettina Silber is Director of Government Relations of Americans For Energy Independence.

Ms. Silber has had extensive experience on Capitol Hill and in public action organizations in the areas of political and international affairs. She served as a Professional Staff Member of the Subcommittee on National Security and the Permanent Subcommittee on Investigations of the Senate from 1972–76, under the Chairmanship of Senator Henry M. Jackson. She has also been Research Director for the American Israel Public Affairs Committee and an editorial writer for the *Near East Report*, a Washington newsletter.

Ms. Silber is a graduate of Goucher College and holds an M.A. in International Relations from the University of Pennsylvania.

Clarice R. Feldman is General Counsel of Americans For Energy Independence.

Ms. Feldman has been a practicing attorney for over 14 years since graduating in 1965 from the University of Wisconsin Law School. She was an appellate attorney for the National Labor Relations Board from 1965 through 1969 when she left to act as co-counsel to Joseph A. Yablonski, candidate for the presidency of the United Mine Workers of America. When Mr. Yablonski, his wife and daughter were murdered, Ms. Feldman joined with his son, Joseph A. "Chip" Yablonski, to represent the reform wing of the UMWA. After the reformers won in 1972, Ms. Feldman served as Associate General Counsel of the UMWA. Subsequently she entered private practice in Washington.

Ms. Feldman is a graduate of the University of Wisconsin.

About Americans For Energy Independence

Americans For Energy Independence (AFEI) is a nonprofit coalition uniting members of the business, labor, academic, scientific, industrial, consumer, conservation, ethnic and religious communities throughout the nation in pursuit of effective energy policies that promote economic growth, social opportunity, and national security.

OFFICERS

Chairman of the Board	Dr. Hans A. Bethe
President	Mr. Joseph D. Keenan
Vice Chairman of the Board	Dr. Zalman M. Shapiro
Secretary	Mr. Harold Greenwald
Treasurer	Hon. Endicott Peabody
Chairman, Executive Committee	Mr. Robert R. Nathan

BOARD OF DIRECTORS

PROF. HANS A. BETHE, Nobel Laureate
Cornell University

MR. OTES BENNETT, President
North American Coal Company

DR. ELIHU BERGMAN, Executive Director
Americans For Energy Independence

RIGHT REVEREND JOHN H. BURT
Episcopal Bishop of Ohio

MR. EDWARD CARLOUGH, President
Sheet Metal Workers International Union

MR. EDWARD E. CARLSON, Chairman
United Airlines, Inc.

MR. THOMAS R. CLARK, Manager
Washington Region
General Electric Co.

MS. EVELYN DUBROW, Vice President
International Ladies Garment Workers Union

MR. HAROLD GREENWALD
Danziger, Bangser, Klipstein,
  Goldsmith & Greenwald

MR. GERALD GRINSTEIN
Preston, Thorgrimson, Ellis & Holman

MR. JOSEPH D. KEENAN, Past
  International Secretary
International Brotherhood of Electrical
  Workers

HON. JOHN NASSIKAS
Squire, Sanders and Dempsey
Former Chairman, Federal Power Commission

MR. ROBERT R. NATHAN
Consulting Economist

MS. LAUREL PARKER
Director of Association Activities
Southern California Edison Co.

HON. ENDICOTT PEABODY
Peabody, Rivlin, Lambert & Meyers
Former Governor of Massachusetts

MR. JAMES T. RAMEY, Vice President
Stone & Webster Engineering Corp.

PROF. ROGER REVELLE, Program in Science &
  Public Policy
University of California, San Diego

MR. BAYARD RUSTIN, President
A. Phillip Randolph Institute

DR. ALBERT SABIN
Research Professor of Biomedicine
Medical University of South Carolina

DR. ARNOLD E. SAFER
Resource Planning Associates, Inc.

DR. ZALMAN M. SHAPIRO, Manager
Fusion Power Systems
Westinghouse Electric Corp.

DR. JOSEPH STERNSTEIN, Rabbi
Temple Beth Sholom, Long Island

DR. DANIEL THURSZ, Executive Vice President
B'nai B'rith International

MR. J. C. TURNER, President
International Union of Operating Engineers

MR. WILLIAM J. VAN NESS, JR.
Van Ness, Feldman and Sutcliffe

MR. MARTIN WARD, President
United Association of Journeymen & Apprentices of
the Plumbing and Pipefitting Industry

MS. MARGARET BUSH WILSON
Chairman, NAACP

STAFF

Executive Director	Dr. Elihu Bergman
General Counsel	Ms. Clarice Feldman
Technical Director	Mr. Merrill Whitman
Director of Government Relations	Ms. Bettina Silber
Director of Communications	Ms. Sherry Saunders
Director of Community Programs	Ms. Shirley Sutton
Executive Assistant	Ms. Britt W. David

ADVISORY COUNCIL

MR. A. J. ALEXANDER, Vice President
Continental Oil Company

DR. S. G. BANKOFF, Chairman
Energy Engineering Council
Northwestern University

MR. ROBERT BEREN, Co-Chairman
Small Producers For Energy Independence

PROF. BERNARD COHEN, Director
Scaife Nuclear Physics Laboratory
University of Pittsburgh

MR. RALPH B. DEWEY, Assistant to the Chairman
Pacific Gas and Electric Company

MR. BERNARD FALK, President
National Electrical Manufacturers Association

DR. FRANK GANNON, Advisor to the Secretary
General
Organization of American States

MR. GRENVILLE GARSIDE
Van Ness, Feldman & Sutcliffe

DR. MELVIN B. GOTTLIEB, Director
Princeton Plasma Physics Laboratory
Princeton University

DR. LAWRENCE GOLDMUNTZ
Economic and Scientific Planning

MS. CAMILLE HANEY, Executive Director
Consumer Concepts

FATHER THEODORE HESBURGH, C.S.C.
President
Nortre Dame University

PROF. WALTER R. HIBBARD, JR.
Virginia Polytechnic Institute and State University

PROF. ROBERT HOFSTADTER
Nobel Laureate
Stanford University

PROF. GORDON M. KAUFMAN
Alfred P. Sloan School of Management
Massachusetts Institute of Technology

MR. RICHARD KLINE, Executive Director
Council of Active Independent Oil and Gas
Producers

MR. ROBERT LEHRMAN, Executive Vice President
American Forest Institute

MR. SIDNEY M. LEVESON, Chief Economist
Gordian Associates, Inc.

MR. JEROME LEVINSON, General Counsel
Inter-American Development Bank

DR. ABRAHAM LILLIENFELD
School of Public Health
Johns Hopkins University

DR. HENRY LINDEN, President
Gas Research Institute

PROF. SEYMOUR MARTIN LIPSET
Hoover Institution
Stanford University

DR. ROBERT MARSHAK
Virginia Polytechnic Institute and
State University

DR. MARGARET MAXEY
Assistant Director
Energy Research Institute

PROF. ADEN MEINEL
MS. MARJORIE MEINEL
Optical Sciences Center
University of Arizona

PROF. H. P. MEISSNER
Department of Chemical Engineering
Massachusetts Institute of Technology

MR. ROBERT PARTRIDGE, Executive Vice President
National Rural Electric Cooperative Association

MR. RICHARD B. POOL, Corporate Energy
   Coordinator
Kaiser Aluminum & Chemical Corporation

DR. WESLEY POSVAR, Chancellor
University of Pittsburgh

MR. WOODRUFF PRICE, Vice President
Seaboard Coast Line Industries

MR. ALEX RADIN, Executive Director
American Public Power Association

PROF. MAURICE SHAPIRO
School of Public Health
University of Pittsburgh

MR. MILTON SHAW
Energy Consultant

MR. JIM SHEETS, Research Director
Laborers' International Union of North America

DR. PETER SHEN, Dean
Hanford Graduate Center

MS. CANDICE J. SHY
Director, Federal Relations, Enserch Corporation

HON. STEVEN SKLAR
Maryland House of Delegates

MR. WALTER P. STERN
Vice Chairman, Capital Research Company

HON. JOSEPH SWIDLER
Leva, Hawes, Symington, Martin & Oppenheimer
Former Chairman, Federal Power Commission

MOST REVEREND ERNEST L. UNTERKOEFLER
Roman Catholic Bishop of Charleston, S.C.

MR. CLARKE WATSON, Chairman
American Association of Blacks in Energy

DR. MAX L. WILLIAMS, Dean
Department of Engineering
University of Pittsburgh

DR. WARREN WITZIG, Chairman
Department of Nuclear Engineering
Pennsylvania State University

DR. GERALD YONAS, Manager, Fusion Research
Sandia Laboratory

MR. H. J. YOUNG, Senior Vice President
Edison Electric Institute

MR. CHARLES YULISH
Charles Yulish Associates

Other Volumes in the AFEI Energy Policy Series

New Perspectives on the International Oil Supply
  Edited by Bettina Silber, Elihu Bergman   1979

An authoritative statement on neglected opportunities to increase and diversify global oil production and break OPEC's monopoly over the pricing and flow of world oil supplies; includes presentations by Senator Edward M. Kennedy and Congressman Charles Vanik.

Alaskan Energy Potential and Constraints to Development
  Edited by Clarice R. Feldman, Bettina Silber   1979

A comprehensive analysis of the role Alaska's resources can play in reducing U.S. dependence on foreign oil and the legislative and administrative climate which best balances environmental concerns and energy needs; includes presentations by Secretary of the Interior Cecil D. Andrus and Senators Henry M. Jackson and Ted Stevens.